Wilderness Warriors

Survival Skills For Kids

A. Hunter

Disclaimer
The information in this book is provided for educational and entertainment purposes only. While every effort has been made to ensure accuracy and completeness, no warranties—express or implied—are given. The author and publisher assume no responsibility for errors, omissions, or outcomes related to using this material. For legal, medical, financial, or other professional advice, please consult a qualified expert.

Safety and Liability Notice

This book contains survival activities and skills designed for educational and entertainment purposes. All activities should be practiced under the direct supervision of a responsible adult.

By reading and using the information in this book, readers and their supervising adults acknowledge that survival skills carry inherent risks. The author and publisher are not responsible for any injuries, accidents, losses, or damages that may result from practicing or applying these skills. Always use common sense, practice safety first, and seek professional guidance if you are uncertain about any activity or situation.

Remember: Wilderness Warriors stay safe, smart, and supervised!

Contents

Introduction

Welcome to the Wild!

Did you know that an adult can only last about three days without water before becoming dangerously dehydrated? Or that you lose more heat from your head than anywhere else on your body? That's right—your mom was right about wearing a hat in the winter!

Survival isn't just for lost explorers or reality TV stars—it's for anyone who loves the outdoors, wants to be independent, and isn't afraid of a little dirt under their fingernails.

Who Is This Book For?

If you love adventure, this book is for you! Whether you're a curious kid, an outdoor explorer, or someone who just wants to be prepared, these survival skills will give you the confidence to handle the wild like a true Wilderness Warrior.

No matter your experience level, if you're ready to learn, explore, and get your hands a little dirty—let's dive in!

Why Learn Survival Skills?

Imagine this:

You're on a camping trip with your family, but somehow, you get separated on the trail.

Your phone? No service.

Your snack? Already eaten.

The sun is getting lower, and you start to feel that uh-oh feeling in your stomach.

Do you:

A) Panic and run in circles until you're dizzy?

B) Cry loudly and hope a friendly bear adopts you?

C) Take a deep breath, remember your training, and use your survival skills to stay safe until help arrives?

If you picked C, you're already thinking like a Wilderness Warrior!

What You'll Learn in This Book

This book is packed with essential survival skills that will turn you into a true Wilderness Warrior. Here are some of the skills you'll master:

1. **Safety** – Assess risks, avoid dangers, and be prepared for anything.
2. **Building a Shelter** – Stay warm, dry, and protected in the wild.
3. **Starting a Fire** – A must-have skill for warmth, cooking, and survival.
4. **Finding Water** – Locate safe drinking water and avoid dangerous sources.
5. **Weather Survival** – Watch for storms, stay dry, and handle extreme conditions.
6. **Foraging for Food** – Identify what's safe to eat and what to avoid.
7. **Navigation** – Read the land, use a compass, and find your way.

8. **First Aid** – Treat scrapes, bites, and minor injuries like a pro.
9. **Campfire Cooking** – Become a wilderness chef with tasty outdoor meals.

Then, put your survival skills to the test with a Final Wilderness Warrior Challenge!

Are You Ready?

By the time you finish this book, you won't just know survival skills—you'll feel ready for any outdoor challenge that comes your way.

So grab your gear, step outside, and let's start this adventure together!

Before You Begin: Stay Safe & Be Smart

Wilderness survival skills are awesome to learn and can help you stay safe in the great outdoors, but remember—you should never intentionally get lost or put yourself in danger just to practice!

- Always learn and practice these skills with an adult.
- Some survival activities involve fire, sharp tools, and outdoor risks. Be smart and stay safe.
- These skills are useful for emergencies and outdoor fun, but the best survival skill is avoiding danger in the first place.

By practicing wisely, you'll become a **Wilderness Warrior**—ready for adventure while staying safe!

Let's Go!

Survival Skill 1: Safety

"Survival can be summed up in three words—never give up." — Bear Grylls

The Most Important Survival Skill: Staying Safe

Before you start building shelters, making fires, or foraging for food, there's one survival skill that's more important than all the rest: **staying safe in the first place!**

When you're out exploring, your number one goal is to return home in one piece. It doesn't matter if you're deep in the forest, walking through your neighborhood, or just hanging out at the park—danger doesn't always come with warning signs. Sometimes, it sneaks up on you like a wolf in sheep's clothing, looking harmless until—BAM!—you're stuck in a bad situation.

Let's test your survival smarts.

Imagine this:

"You're hiking on a crisp fall afternoon, following a well-worn path through the woods. Up ahead, a narrow side trail catches your eye. It looks like a shortcut. No big deal, right? You take it. But after a few turns, the trees start blurring together. The trail is gone. Your stomach sinks. You're lost."

Pause and think for a moment—what do you do?

A) Walk in the direction you think is right and hope for the best.

B) Panic, run, and yell (because that always helps... right?).

C) Stop, stay calm, and look for clues to help you get back.

If you picked **C, nice work!** Staying put is the smartest move. Many people make things worse by wandering deeper into the unknown. Instead, take deep breaths, use a whistle if you have one, and look for familiar landmarks. If you're near a trail, listen for voices. Want to be found fast? **Make it easy for rescuers**—staying put is often the smartest move!

The same goes for everyday dangers. If you see a puddle of murky brown water, do you take a sip? (Hopefully not—yuck!)

If you find a frozen lake, do you charge onto the ice like a stampeding moose, or do you test it first?

If the weather is freezing or scorching hot, do you just tough it out, or do you gear up?

Exactly! You wouldn't drink weird water (unless you enjoy stomachaches), you'd check the ice before stepping on it, and you'd dress smart to survive extreme weather.

The best way to stay safe? Be prepared. Learning survival skills now will help you recognize dangers before they happen—whether it's spotting risky situations, knowing how to handle emergencies, or avoiding common mistakes that can turn an adventure into a disaster.

The 10 Golden Safety Rules

A lot of people think the scariest thing in the wild is a bear. But let's be real—getting lost, running out of water, or underestimating the weather is way more likely to ruin your adventure. That's why every great Wilderness Warrior should follow these 10 Golden Safety Rules:

1. **Never trust a stranger just because they seem nice.** Not everyone is who they say they are. If someone you don't know tells you to go with them, don't do it—even if they say they know your family. Instead, have a secret password that only your trusted adults know. No password? No going!

2. **Stick with a buddy.** Wandering off alone is a great way to end up lost, and phone signals aren't always reliable in the wild. Always tell someone where you're going, and take a buddy along for the adventure!

3. **Pack the right gear.** The weather can change fast, even if your coat says otherwise. Bring a survival kit with essentials like food, water, extra clothes, and navigation tools. Oh, and don't forget a first aid kit—you never know when you'll need it!

4. **Plan ahead.** A smart adventurer never just "wings it." Plan your trip with a trusted adult, check the weather, and make sure you have everything you need. Think like an adventure guide—always be prepared!

5. **Leave no trace.** Nature isn't your personal trash can. Keep campsites clean, don't chop down trees, and be mindful of fires. If you brought it with you, take it

back—try to leave everything just as you found it, except for great memories. Those are yours to keep!

6. **Stay on the trail.** Trails exist for a reason: to keep you safe and help you avoid getting lost. Don't be that person who ignores the path and ends up starring in a real-life survival story.

7. **Wild animals are not your friends.** A sick or injured animal might seem like it needs help, but getting too close could be dangerous—it might have mites, rabies, or lash out in fear. The best thing to do? Keep your distance! If you think an animal truly needs help, contact a local wildlife rescue, animal control, or park service.

8. **Drink water—lots of it!** If you're feeling thirsty, you're already dehydrated. Sip water throughout the day, and if you're out in the wild, make sure it's purified before drinking. You can't see, smell, or taste harmful bacteria—and while some bacteria are good, the wrong kind can make you seriously sick (you don't want that—bleh!). Stay safe and purify your water first!

9. **Bring a first aid kit.** A tiny cut can turn into a big problem if you don't take care of it. Pack bandages, antiseptic wipes, and any personal medications you might

need. Bonus points if you know how to use them!

10. **Respect the sun.** Sunburns aren't just painful—they zap your energy and make dehydration worse. Wear a hat, cover up, or slap on some sunscreen (yes, even when it's cloudy). Your skin will thank you!

What to Do if You Get Lost

First rule of the wild? **Don't wander off alone!** Always stick with a buddy or an adult because getting lost isn't as fun as it sounds in adventure movies. But let's say, somehow, you look up, and—uh-oh—you have no clue where you are. Your stomach drops like you just missed the last step on a staircase, and suddenly, you're sweating like a popsicle in the sun. That's totally normal! But here's the key: **Don't panic.**

The initial shock can make you feel scared—and that's completely normal. But panicking? That's the real danger. If you stay calm, you can think clearly, and guess what? You have everything you need to handle this. But, that's why you should always carry an **emergency kit**—because smart adventurers (like you) come prepared. Let's check what's inside!

The Ultimate Lost-in-the-Wild Survival Kit

Here's what every Wilderness Warrior should have in their pack:

- **Snacks!** Two Ziplock bags: one with a sandwich (eat this first) and another with something energy-packed, like trail mix. And don't toss those bags—they might just help you collect water later!

- **A Signal Flag.** Tie something bright (like an orange plastic bag) to a branch near you. The movement helps searchers spot you.

- **A Whistle.** Yelling for help drains your energy fast, but a whistle takes barely any effort. Three short blasts mean **SOS**—the universal "Help me!" signal.

- **A Reflector.** A mirror or piece of foil (yes, the inside of a snack bar wrapper will work) wrapped around a stick works great! Catch some sunlight and flash it toward people or planes.

- **A Waterproof Poncho.** Not only does it keep you dry, but if it's big enough, it doubles as a mini tent. Bonus points if it's bright yellow or orange—it makes you easier to find!

Why This Gear Matters

Having food and shelter lets you **stay put** instead of wandering. And staying put means getting found faster. Find a big tree and stick to it like glue.

Staying Safe While You Wait

- **Stay warm.** Keep all your clothes on (unless they're wet) and avoid sitting right on the ground—it steals your body heat fast. Use leaves, branches, or your bag as insulation.

- **Don't hide!** Pick a spot out of the wind and rain, but stay visible. Hiding in a bush might sound nice—until no one can find you!

- **DO NOT eat mystery berries.** Or mushrooms. Or anything you don't 100% recognize. We'll learn safe foraging later, but for now? If you don't know it, **don't eat it.**

- **Avoid big rivers and lakes.** Falling in means you'll be wet, cold, and even more lost. If you need water, stick to small, slow-moving sources and use your plastic bag to collect it. Tell an adult you drank from it when they find you.

- **Blow your whistle.** Every few minutes, give three short bursts—whistle sounds travel farther than shouting. Then rest. Keep this up—it makes it much easier for rescuers to track you down!

- **Scared? Sing a song!** It sounds silly, but singing helps you feel braver. Plus, if rescuers hear you, they'll come running.

Water Safety: Fun but Fierce

Water is awesome—until it's not. Even if you're the best swimmer in town, never go into water alone. That goes for rivers, pools, lakes, and oceans. Here's how to stay safe:

- **A lifeguard or adult should always be watching**. If they have to leave for even just a moment, wait until they come back to continue your underwater dives.

- **Protect your feet.** Sand and mud can hide broken glass or sharp rocks, so wear shoes!

- **Beware of underwater plants.** If seaweed or weeds tangle around your legs, don't thrash—it only makes things worse. Stay calm, move slowly, and if needed, wave one hand for help.

- **Wear a life jacket.** While swimming, you may or may not need one, but if you're on a boat, jet ski, or any kind of watercraft, you absolutely need one. No exceptions! If you fall in, it helps you float, and its bright color makes you easy to spot.

- **Watch out for sneaky ocean currents**. Some are obvious, but rip currents? Not so much. Before swimming in the ocean, ask a lifeguard where they are and how to avoid them.

"Pro Tip: Protect your body from the sun—wear lightweight clothes or slather on sunscreen before and after swimming. Sunburns are the worst kind of survival challenge!

Let's test your water smarts.

Imagine this:

You and your friends are hanging out by a river. You see a cooler floating by, looking like it's just waiting to be pulled in. You reach out to grab it, but the current pulls you off balance, and suddenly—whoosh!—you slip and fall in. The water's icy, and the current is much stronger than you thought. What should you do?

Here's what you need to know:

- **Stay calm!** Panicking can make everything worse.
- **Float on your back.** This helps you stay above the water and keep your breathing steady.
- **Move sideways.** Instead of fighting the current head-on, try to move to the side to escape the strongest flow.
- **Never enter fast-moving water alone.** Always make sure someone is with you, and never go near rivers or streams that look too dangerous.

Final Thought

The wilderness is an amazing place—but staying safe means being smart. Follow these tips, and you'll be a true Wilderness Warrior ready for any adventure!

Activity 1: Safety Quiz!

Let's test your survival smarts with a true-or-false quiz! For each statement, decide if it's true or false—and explain why. Sometimes, there might be more than one correct way to think about it.

Ready? Let's go!

1. Chasing a wild animal in the woods is a fun way to practice survival skills.

False.
Wild animals should be left alone. Chasing them could lead you off the trail, get you lost, or even put you in danger if the animal feels threatened.

2. If a stranger asks you for directions, it's okay to help them.

False.
If a stranger needs help, it's safest to find a trusted adult instead of talking to them yourself. Stay cautious and never go anywhere with someone you don't know.

3. As soon as you arrive at camp, you should go exploring alone.

False.

Never wander off alone! Always stay with a buddy or a trusted adult so you don't get lost or run into danger.

4. It's safe to run ahead of your group while hiking.

False.

If you get too far ahead, you might take a wrong turn, get hurt, or be out of sight when you need help. Stay close and stick together.

5. A phone is the most important thing in your survival kit.

False.

While a phone can be helpful, it might not always work. Other survival tools like a whistle, poncho, flashlight, and first aid kit can be life-saving if you're lost.

6. If you see trash on the trail, you should pick it up and throw it away.

True.

Following the "Leave No Trace" rule helps keep nature clean. Just be careful—use gloves or a bag if the trash looks unsafe to touch.

7. When hiking, you should always stay on marked trails.

True.

Staying on the trail helps prevent getting lost, keeps you away from dangerous areas, and protects the environment.

8. You should only drink water when you feel thirsty.

False.

If you wait until you're thirsty, you're already dehydrated! Sip water regularly. If you're lost and have a limited supply, take small sips and look for safe water sources.

9. A first aid kit is optional when camping.

False.

A first aid kit is a must-have! It should include bandages, disinfectant, and any necessary medications in case of an injury.

10. The correct emergency whistle signal is one long blast.

False.

The universal SOS signal is three short whistle blasts—this tells rescuers you need help. Bonus points if you can demonstrate it!

Bonus Challenge!

Think you can come up with more safety questions? If you're in a group, write down a few true-or-false questions, put them in a hat, and take turns drawing and answering them

How Did You Do?

Count the number of correct answers you got from the quiz, and then see what level of safety expert you are!

- **Safety Starter (0–3 Correct):** You're just beginning your journey to becoming a safety expert. Keep reviewing the rules and try the quiz again—every great Wilderness Warrior started right here!

- **Safety Specialist (4–7 Correct):** Nice work! You've got a solid grasp on the safety tips. With a bit more practice, you'll be ready to handle any adventure with confidence.

- **Safety Master (8–10 Correct):** Wow, you're a true expert! Your quick thinking and careful attention to safety make you a real Wilderness Warrior. Keep up the amazing work!

Remember, the goal is to learn and have fun. Even if you don't score as high as you'd like, you can always study the tips again and try another quiz later!

Activity 2: Nature Scavenger Hunt

Ready to explore like a true Wilderness Warrior?

Nature is full of surprises—if you know where to look.

Your mission: find the items below, but remember—leave no trace.

Don't touch—just snap a photo, check it off, or sketch it in your journal!

☐ **A bird** (bonus points if it's singing!)

☐ **Some animal tracks** (who left them?)

☐ **A bug** (the weirder, the better!)

☐ **A nest of any kind** (but don't get too close!)

☐ **Three different types of leaves** (can you spot the difference?)

☐ **Some water** (be creative—does a puddle count?)

☐ **A flower** (wild or planted?)

☐ **A fallen branch** (but watch out for sneaky critters underneath!)

☐ **An acorn, pine cone, or seed**

- ☐ **A squirrel or rabbit** (they're quick—stay sharp!)
- ☐ Something you think is **beautiful**
- ☐ Something you think is **surprising**

This isn't a race—it's an adventure! Take your time, look closely, and see if you can spot something most people would miss. Who knows what amazing things you'll discover?

How Did You Do?

Count how many items you found and see what level you reached! Remember, not every environment will have every single object. So, do your best and have fun!

Beginner Explorer (1-5 found) – You're just starting your wilderness journey. Keep your eyes open, and next time, try to spot even more cool details!

Intermediate Adventurer (6-8 found) – Nice work! You're doing great. Sometimes you might not find every single item in your area, but you're really getting the hang of spotting nature's hidden treasures.

Wilderness Warrior (9-12 found!) – Whoa! You have the sharp eyes of a true explorer. You're noticing things that most people miss. Keep it up!

Activity 3: Name That Tree

Are you ready to become a nature detective? In this activity, if you have a smartphone or tablet, you'll use it (with a plant identification app!) to help you identify trees and learn their names. Try downloading an app like vTree, Seek, or another similar tool. There are plenty of options out there, so if one doesn't work for you, there might be an alternative you like. These clever apps can help you figure out which tree you're looking at just by taking a picture!

*If you don't have access to an app or the internet, don't worry! Your local library likely has field guides or books about trees in your area. These guides are especially helpful because they focus on the plants that grow near you.

To help you record your findings, consider starting a nature journal. If you're looking for ideas, check out Nature Journal for Kids: Learn. Explore. Record. (written by me!). It's a great place to keep sketches, rubbings, and photos of your nature discoveries.

How to Play:

1. **Observe the Tree:** Look closely at the tree you've chosen. What do you notice? Check out its leaves, flowers,

fruits, and bark. These details are what experts use to tell trees apart.

2. **Record Your Findings:** In your notebook or nature journal, draw a picture of the tree's leaf (or even the whole tree!) and write down everything you observe.

3. **Snap a Photo:** If you're using an app, take a photo of the tree or its leaves, following the app's instructions.

4. **Get the Name:** Once the app (or your field guide) helps you identify the tree, write down its name in your journal.

5. **Repeat the Fun:** Try this with at least three different trees. You can even add extra sketches or notes about what makes each tree unique. Just remember: if you collect anything like a fallen leaf, always ask for permission—and never pick things directly from a live tree!

How Did You Do?

- **Nature Sleuth (detective):** Found three leaves but have no idea what they are.

- **Junior botanist (plant scientist):** Found at least three leaves and identified them.

- **Expert Tree Detective:** Found at least three leaves, identified them, and wrote down interesting facts about each one.

Wilderness Warrior Skill Unlocked!

You've mastered the art of safety! Now, mark your progress by coloring in your Wilderness Warrior Safety Award on your Survivalist Skill Tracker (page 277).

Level Up Your Survival Skills!

You're officially on your way to becoming a Wilderness Warrior! Now that you know how to stay safe in the wild, it's time for your next survival skill—building a shelter!

Up next: Skill 2 – Creating the Ultimate Wilderness Shelter! Let's GO!

Survival Skill 2:
Building a Shelter

"You can last three weeks without food, three days without water, but only three hours without proper shelter in extreme conditions." — The Rule of Three

Why Shelters Are Essential

Imagine you're out on a great outdoor adventure. Maybe you're headed deep into the woods for some remote camping, out on a long hike, or—oops—maybe you find yourself lost. Suddenly, the weather turns. The temperature drops, the wind picks up, and dark clouds roll in. What do you do?

You build a shelter, of course! Remember the "rule of three"? It tells us that without food you can last three weeks, without water you can survive only about three days, and without shelter in very hot or very cold conditions, you might only last a few hours! And that's why having a safe, cozy shelter is absolutely essential.

When you're facing extreme temperatures, a good shelter can mean the difference between staying warm and dry or feeling frozen and soaked. And here's something important: being cold is actually more dangerous than being hungry! Your body can lose heat really fast, which can lead to hypothermia (that's another word for "too cold to function"). In hot weather, a bad sunburn or dehydration can be just as serious. But fear not—by the end of this chapter, you'll know exactly how to build a shelter that keeps you safe and comfortable!

Two main types of shelters:

- **Short-term shelters:** Perfect for when you need a quick break from the elements.
- **Long-term shelters:** Built when you need a safe, secure place to stay for a longer time.

No matter which type you need, every shelter must have three key features:

- **Insulation:** Keeps you at just the right temperature—warm in the cold and cool in the heat (kind of like those fancy insulated cups that keep your drink perfect).
- **Waterproofing:** Staying dry is essential, because being soaked and shivering is no fun at all.
- **Windproofing:** Blocks the wind from stealing your warmth (that pesky wind chill can be a real bummer).

If your shelter is missing any of these, you risk exposure—getting too cold, too wet, or too windy—which can make you sick and miserable. And that's definitely something you want to avoid!

But here's the catch: you've got to think fast! For example, if it starts to rain, it's no use trying to build a shelter in the downpour. Instead, look for a natural cover—maybe under a big tree, at the entrance of a cave (but keep an eye out for animals!), or beneath thick bushes.

The Best Place to Build Your Shelter

Not every spot is a good spot for a shelter—some places are downright terrible (like a dry riverbed, for example!). Before you start stacking sticks, follow these Golden Shelter Rules to pick the perfect location.

Places to Avoid:

- **Low Areas:** Cold air and water tend to pool here, increasing the risk of flooding.
- **Under Dead Trees or Loose Rocks:** These spots can be dangerous—falling debris or shifting ground is not what you want.
- **Too Close to Water:** While you need water nearby, building right next to it can lead to unexpected floods.
- **Bug Hotspots:** Unless you enjoy having mosquitoes as uninvited roommates, steer clear of areas that attract lots of bugs.

Places to Choose:

- **Flat Ground:** This helps keep your shelter stable—no rolling away while you sleep!
- **Somewhere Dry:** Avoid areas where puddles or rainwater can quickly soak you.
- **Near Natural Materials:** Being close to branches, leaves, and other supplies means you won't have to wander far when building your shelter.
- **Visible for Rescue (but with Some Cover)**: Pick a spot that can be easily found if you need help, yet offers enough natural cover to keep you hidden from potential dangers.

More Tips for Choosing Your Shelter Location:

- **Plan Ahead:** Start building early in the day to ensure you have enough time to set up before it gets dark.
- **Scout Carefully:** Look around to find a safe, flat area with good visibility.
- **Mind the Wind:** The direction of the wind affects

your shelter's warmth. Position the opening of your shelter away from strong winds.

- **Watch for Fire Risks:** If your shelter is made of natural materials like branches and leaves, make sure that smoke or stray embers can't blow inside and start a fire.

- **Consider Elevation:** Avoid building in dips or low spots that might flood or trap cold air. If you must build on a slope, try to level it out or create channels to divert rainwater.

- **Stay Alert to Danger:** Avoid areas under cliffs, near dead trees, or where falling branches could be a risk. If you spot animal tracks (from mountain lions, bears, etc.), it might be best to choose another spot.

- **Proximity to Water:** Aim to build about 200 feet away from a water source. This keeps you close enough to get water, yet far enough to avoid flooding or contamination.

- **Stay Close to a Trail:** This makes it easier for rescuers or friends to find you, but be sure not to build directly on the trail where you might get accidentally stepped on.

By following these guidelines, you can choose a shelter spot that keeps you safe, warm, and dry. Now, get out there and build a shelter that can stand up to the weather!

Before We Build

There's no one-size-fits-all survival shelter. The best type depends on where you are, the materials you have, and how long you need to stay safe. Before you start building, remember that in nature you might not find the perfect pieces of wood—so you may need some tools. And always follow the leave-no-trace rule: practice with fallen branches rather than cutting down live ones unless you're in a true survival situation!

Tools & Preparation

Before you get started, you might need:

- An ax or hatchet
- A sharp, quality knife
- A folding saw
- A small shovel
- A tarp

- A rope or paracord

*All of these tools, when used improperly, can hurt you! Please use them carefully and with an adult.

Ground Insulation

It might be tempting to jump right into building a shelter so you have something over your head, but hold up—before you start stacking sticks, you need to insulate yourself from the ground first. Why? Because the ground absorbs the temperature around it, which means it can be scorching hot during the day or freezing cold at night. If you sleep directly on it, you'll either be sweating buckets or shivering like a popsicle. No thanks!

Step 1: Clear the Area

Grab a branch with some leaves on it—yes, like a broom—and sweep away debris where you plan to build. Trust me, you do not want to sleep on rotting leaves or an infestation of angry creepy crawlies.

Step 2: Build Your Insulation Layer

Now, gather a huge pile of fresh or dry leaves—enough to make a thick, fluffy mattress. Lay on it. If you can still feel the cold, keep adding more. Think Princess and the Pea—but instead of a tiny pea, it's the freezing ground, and you really don't want to feel it. Once it's thick enough, use some branches to create a frame around the edges to keep everything in place.

Pro Tip: Make your insulation layer before you start building shelters like an A-frame or debris shelter. If you're making a lean-to or tarp shelter, you can add insulation afterward. A well-insulated bed will keep you warm, comfortable, and ready for a good night's sleep—because surviving the wilderness is way harder when you're running on zero sleep!

Shelter Types

1. Lean-To Shelter – Quick and Easy

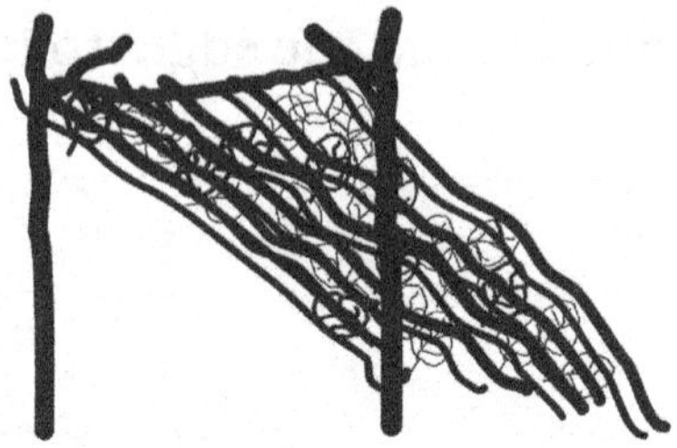

Description: An open shelter with a slanted roof that leans against a sturdy support.

Best for: Quick protection from wind and rain, short-term survival

What You Need:

- Two sturdy Y-shaped branches or two trees close together
- A long, sturdy branch (ridge pole)
- Natural materials like branches and leaves
- Rope
- (Optional) A tarp and a couple stones

How to Build It:

1. **Find Your Support:** Locate two strong trees (or plant Y-shaped sticks in the ground) slightly farther apart than your height—because you'll need room to fit inside!

2. **Set Up the Ridge Pole:** Tie a sturdy branch between the trees or Y-sticks at about hip height. Keeping the shelter small and snug helps trap warmth.

3. **Add a Roof:**

    - **Using a tarp?** Tie it over the ridge pole and angle it down to form a sloped wall. Secure the lower corners with stones.

    - **No tarp?** Gather thick branches (called "ribs") and lean them against the ridge pole to create a sloping wall.

4. **Fill in the Gaps:** Weave smaller branches with leaves into the gaps between the ribs to block wind and rain.

5. **Insulate the Ground:** If you haven't already, add your ground insulation inside the shelter to stay warm and comfortable.

2. A-Frame Shelter – Cozy & Warm

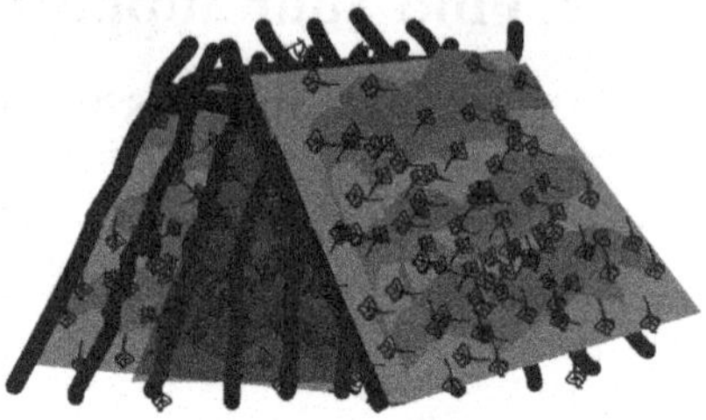

Description: Think of an A-Frame like a lean-to with an extra wall, making it look like the letter "A"! This design traps heat better, keeping you warmer in cold weather. It's like building your own mini log cabin out of branches and leaves!

Best for: Cold weather and better insulation.

What You Need:

- Two sturdy Y-shaped branches or two trees close together
- A long, sturdy branch (ridge pole)
- Natural materials like branches and leaves
- Rope
- (Optional) A tarp and a few stones

How to Build It:

1. **Start with Ground Insulation:** Lay down a thick bed of leaves or other soft materials to keep the cold ground from stealing your warmth.

2. **Build the First Wall:** Follow the Lean-To Shelter steps to create a slanted wall using a tarp or stacked branches.

3. **Add the Second Wall:** Now, build another slanted wall on the opposite side, so both sides lean against the ridge pole—just like the shape of the letter "A"!

4. **Seal the Gaps:** Pile on extra leaves or natural materials to block wind and trap heat. The thicker the walls, the toastier you'll be!

Pro Tip: This shelter takes more effort, so work as a team if possible. A helping hand makes the job faster—and more fun!

3. Debris Shelter – The Ultimate Survival Shelter

Description: The Debris Shelter is like an A-frame shelter, but instead of using two ridge poles, it has a single support pole propped against a tree stump or wedged into a V-notch in a tree. It's packed with natural insulation to trap body heat, making it one of the best shelters for cold weather survival. The thick walls help block wind, rain, and snow, keeping you warm and hidden.

Best for: Extreme cold or long-term survival.

What You Need:

- A tree stump or V-notch in a tree (to hold up your shelter)
- A strong, long branch (support pole)
- Lots of natural materials: branches, leaves, and moss
- Rope (optional)
- A tarp (optional)

How to Build It:

1. **Set Up the Support Pole:** Find a tree stump or a V-shaped notch in a tree to hold one end of your support pole. Secure the other end on the ground using rocks.

2. **Add Ground Insulation:** Before building the walls, make a thick mattress of leaves under the support pole.

3. **Build the Walls:** Stack sturdy branches along both sides of the support pole, like making a mini tent. Fill any gaps with smaller branches and leaves to block wind.

4. **Pile On Insulation:** Cover the walls with at least a foot of leaves or moss. A thick layer will help trap heat inside!

5. **Leave an Air Hole:** You need fresh air! Keep a small opening near the entrance. You can partially block it with a bag or extra leaves—but never seal it completely!

Pro Tip: This shelter takes time and effort, so plan ahead if you're in for a cold night. The more leaves you pile on, the warmer you'll be!

4. Tarp Shelter – Fastest Setup

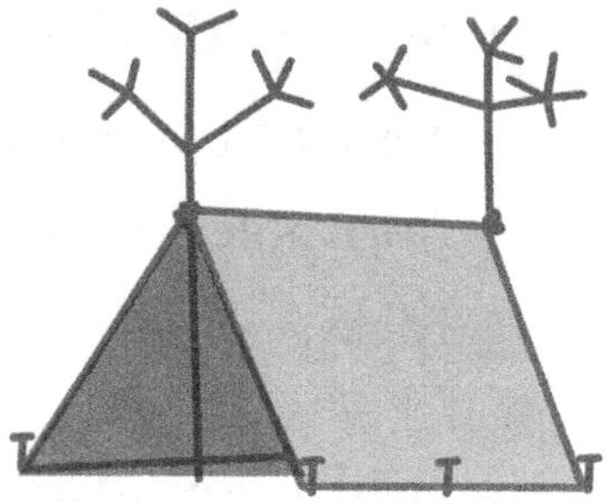

Description: Need shelter in a hurry? The Tarp Shelter is like a portable roof that protects you from rain, wind, and sun in just minutes! It's lightweight, easy to set up, and perfect for short-term survival or unexpected storms. While it won't trap heat like a debris shelter, it keeps you dry and gives instant protection—great for staying comfortable while you set up a more insulated camp.

Best for: Rainy weather, ultralight survival, and quick protection.

What You Need:

- A tarp
- Rope or paracord
- Trees or rocks for anchors
- A tall branch (to prop up the tarp)
- (Optional) A knife (to cut rope)

How to Build It:

1. **Set Up the Roof:** Tie your tarp between two trees to create a simple overhead cover.

2. **Angle for Rain Drainage:** Tilt one side of the tarp toward the ground so rainwater runs off easily instead of pooling.

3. **Secure the Edges:** Tie the corners of the tarp to rocks or sticks to keep it from blowing away.

4. **Prop It Up:** Place a tall branch under the center of the tarp to form a tent-like shape, helping water drain properly.

5. **Add Ground Insulation:** Lay down leaves or other insulation inside for comfort and warmth.

6. Expand for Groups: Need more space? Attach extra tarps to create walls, using branches for support.

Tool Tips:

You won't always have perfect tools in a survival situation, so get creative! If you don't have a knife, use rocks or wedged branches to help break longer sticks or branches into the right size.

Final Thought:

By choosing the right shelter and using these steps, you can stay dry and safe no matter the weather. Whether you're camping under the stars or in a real survival situation, you now have the skills to build a sturdy, reliable shelter!

Activity 1: Wilderness Warrior Shelter Challenge!

Mission: You're stranded in the wild, and night is falling! Your survival depends on building a strong, weatherproof shelter. Think like a true Wilderness Warrior—where will you set up camp?

Step 1: Scout Your Survival Spot

Before you start building, ask yourself:

- Is the ground flat, or will rainwater pool here?
- Are there dangers nearby, like loose rocks or dead tree branches overhead?

- Which way is the wind blowing? You don't want a chilly breeze sneaking in!
- What resources can you use—branches, leaves, moss, or even a tarp?

Step 2: Choose Your Shelter Type

- **Is it warm?** A simple lean-to might do.
- **Cold and windy?** You'll need a debris shelter for insulation.
- **Rainy?** A tarp shelter will keep you dry.

Step 3: Gather Your Supplies

Be creative! Look around and use the resources available. If you have tools like rope, tarps, or a saw, use them wisely. No tools? No problem! Find fallen branches, leaves, and rocks to build your shelter.

Step 4: Build Like a Pro

Start constructing your shelter! If you're in a group, work together like a survival team! If you're on your own, put your skills to the test and see what you can build solo.

Step 5: Shelter Check!

Once your shelter is complete, put it to the test:

- Would it block the wind?
- Is there enough room to fit inside?
- Would rain roll off, or would you wake up in a puddle?

Survival Debrief – Questions to Think About

- Did you have everything you needed? What was missing?
- What was the hardest part of building your shelter? How could you make it easier next time?

Rate Your Shelter: **5 STARS** – Would you sleep in it overnight?

4 STARS – Pretty solid, but a raccoon might steal my spot.

3 STARS – Needs work… I'd probably get wet.

2 STARS – I don't trust it. At all.

1 STAR – Collapsed instantly.

Activity 2: Knot Mastery – The Square Knot Challenge!

A strong shelter needs strong knots! If you want your branches to stay in place and your shelter to last through wind and rain, you'll need to tie them securely. That's where knot-tying skills come in!

The Challenge: Learn the Square Knot

The square knot is one of the most useful survival knots. **It's perfect for:**

- Joining two shorter ropes together to make a longer one.
- Securing branches, tarps, or gear in place.
- Creating a strong hold that's easy to untie when needed.

How to Tie It (The "Right Over Left, Left Over Right" Trick!)

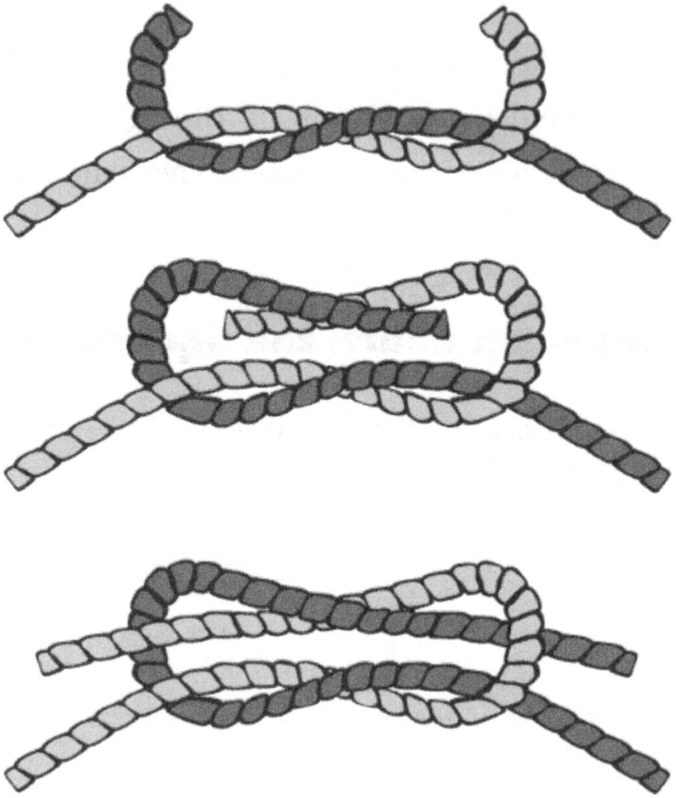

1. Take two ends of the rope (using different colors can help).

2. Cross the **right** rope **over** the left, then tuck it **under**—just like the first step of tying your shoes.

3. Take the **left** rope (now on the right side) and cross it **over** the right, then tuck it **under** again.

4. Pull both ends tight—your knot should lie flat and look symmetrical.

Helpful Hint: If your knot looks uneven, twisted, or slips when pulled, you might have tied a **granny knot** instead. A square knot should hold firm but still be easy to untie. If it doesn't look right, try again!

Put Your Skills to the Test!

Speed Round: How fast can you tie a perfect square knot? Time yourself and see if you can beat your record!

Strength Test: Tie two ropes together and give them a few tugs—does your knot hold strong?

Survival Simulation: Use your knot to attach a branch or secure a tarp as if you were building a real shelter!

Already a Pro?

Want to take your skills to the next level?

Try learning these two essential survival knots!

Taut-Line Hitch – The Adjustable Grip

Perfect for tightening tent lines or securing tarps, the taut-line hitch creates an adjustable loop that grips tight under tension.

Here's How:

1. Wrap the rope around a sturdy object (like a pole or tent stake).

2. Bring the free end around the standing line and loop it through itself twice.

3. Make a third loop, but this time on the outside of the first two.

4. Pull tight. You should be able to slide the knot up and down to adjust tension.

Clove Hitch – Quick and Secure

A great all-purpose knot for securing ropes to trees or poles.

1. Wrap the rope around a pole or branch.

2. Cross over itself and wrap around again.

3. Tuck the end under the second loop and pull tight.

That's it! It should hold strong but be easy to untie when needed!

Pro Tip: Shoe laces or paracord are great for practice! Keep practicing until you can do it with your eyes closed—because in a real survival situation, you might need to tie a knot in the dark!

Activity 3: The Ultimate Waterproof Shelter Challenge

Imagine dark clouds rolling in, the wind picking up, and the first drops of rain splashing on your face. You need shelter—fast! But not just any shelter—it has to be completely waterproof. Will your survival skills keep you dry, or will you end up soaked?

For This Lesson, You'll Need at Least One Other Person!

Survival is a team effort, and this challenge is best done with a partner or a group. One person will build while another tests, observes, and helps troubleshoot. Working together will make your shelters stronger and your skills sharper!

Your Mission

Build two shelters:

1. **A Natural Shelter (Debris, A-frame, Lean-to)** – Using only branches, leaves, and debris.

2. **A Tarp Shelter** – Secured with knots and rope.

Once your shelters are built, it's time for the **Waterproof Test!** One brave survivalist climbs into each shelter while another pours water over it using a watering can. The goal? **Stay as dry as possible!**

Survival Debrief - What Went Wrong?

After the test, gather your team and analyze the results. Ask:

- How well did each shelter hold up against the water?
- What areas leaked the most?
- What improvements could be made for next time?
- If you were in a real survival situation, what would you do differently?

For extra credit, tweak your designs and try again! Can you improve your waterproofing?

Recap: Why Shelter Matters

Without shelter, extreme temperatures can put you in danger within hours. A well-built shelter isn't just about comfort—it's about staying safe, dry, and warm in the wilderness.

Pro Tips:

- **Look around for the best location**—high ground is

better than a low spot that could flood.

- **Use the right materials**—big leaves, bark, or layered branches can help block wind and rain.
- **Test your work**—don't wait for a real storm to find out your shelter has a leak!

Safety First: When using tools like knives or saws, always have an experienced guide with you. Learning how to use tools is an important survival skill, but never use them alone!

Wilderness Warrior Skill Unlocked!

You've mastered the art of shelter-building! Now, mark your progress by coloring in your Wilderness Warrior Safety Award on your Survivalist Skill Tracker (page 277).

Level Up Your Survival Skills!

Now you know how to build a strong shelter and stay protected from the elements. But what if the temperature drops, and you need warmth fast? It's time to master one of the most important survival skills.

Up next: Skill 3 – Starting Fires Like a Pro! Let's GO!

Survival Skill 3:
Starting a Fire

"A spark neglected makes a mighty fire." — Robert Herrick

Mastering Fire – Friend or Foe?

Fire is one of the greatest survival tools ever discovered—but also one of the most dangerous. It can warm your frozen fingers, cook your food, and light up the darkness, but it can also spread out of control, destroy forests, and put lives at risk.

Think of fire like a wild animal. If you train it, it can be helpful and powerful. If you ignore the rules, it can turn into a raging beast. A true Wilderness Warrior respects fire and knows how to use it safely—because if you don't, you might end up like a marshmallow that got too close to the flames... totally charred!

Fire Safety 101: Think You Know the Rules?

You wouldn't ride a bike blindfolded or play baseball with your eyes closed, right? Fire needs just as much caution. Before you learn how to build one, make sure you know the fire safety rules!

Don't:

- **Chop firewood if you're not trained.** Swinging an axe is serious business—leave it to adults.
- **Start a fire wherever you want.** Always use a fire pit or a designated area, or you could start a wildfire—not cool!
- **Run around the fire.** A campfire is not a racetrack. Trip and fall? Say hello to burnt shoes—or worse.
- **Walk between the fire and someone's chair.** The last thing you want is to trip and face-plant into the flames.
- **Let your fire get too big.** If it's taller than three feet, it can become dangerous fast.

Do:

- **Use a safe fire spot.** Stick to fire pits and cleared areas where flames can't spread.
- **Keep fire-fighting tools nearby.** Always have two buckets—one with sand and one with water—to put out flames or stop flying embers.
- **Keep the fire small and manageable.** A controlled fire is a safe fire.
- **Supervise the fire at all times.** If no one is watching it, it shouldn't be burning!

Follow these rules, and your campfire will stay warm, safe, and fun!

*Pro Tip: Plan Ahead!

Check Fire Regulations

Not all campsites allow campfires, and some don't even sell firewood. In hot, dry summers, many national parks and states ban fires entirely to prevent wildfires. Before you head out, help your family research the fire rules for your camping spot. You can check fire restrictions at your local Department of Natural

Resources (DNR) or on the U.S. Forest Service website: www.fs.usda.gov/Internet/FSE_DOCUMENTS/fseprd587410.pdf

Don't Bring Unwanted Stowaways!

Firewood can carry hidden bugs and diseases that could harm forests in new areas. To protect the environment, many parks have strict rules about outside firewood and may require you to buy firewood locally. Before packing your own, check the park's regulations or visit DontMoveFirewood.org for more information.

How to Prevent Wildfires

Like that famous bear says, wildfires can be prevented! In fact, most wildfires start because someone didn't handle their fire with care. Let's go over the golden rules of fire safety so you can help protect the great outdoors.

The Five Fire Rules:

1. **Take responsibility.** If you start a fire, you are in charge of keeping it safe.

2. **Be careful.** Treat fire with respect—it's not a toy.

3. **Never play with lighters or matches**. A small spark can turn into a big disaster.

4. **Watch your fire.** Keep an eye on embers—they can travel and start new fires.

5. **Put it out completely.** Before leaving, make sure the fire pit is cold. If it's warm, it's still a danger!

Follow these five rules, and you'll help protect the forest and all the animals that call it home.

What Do You Need to Build a Fire?

Before you strike a match, you need the right setup. Fire-building is more than just lighting a pile of sticks—it's a process!

First things first: Get permission! Never light a fire without an adult's approval and supervision. Once you have the green light, follow these steps:

- **Find (or make) a fire pit.** If there's no designated fire ring, you'll need to create a safe space.
- **Look up!** Check for dead branches or dry leaves above your fire area. Floating embers can set them ablaze.
- **Clear the ground.** Remove dry leaves, twigs, or anything that could catch fire within a 10-foot circle. Keep all firewood outside this zone.
- **Dig a pit.** Make a 1-foot deep and wide hole in the center of your cleared area.
- **Ring it with rocks.** Find large, sturdy rocks to line your fire pit—but never use river rocks! Over time, they absorb water, and when heated, the trapped moisture turns to steam, building pressure until the rock cracks or even explodes! Instead, use solid, dry rocks like gran-

ite, found away from water sources.

Pro Tip: Reduce Smoke with a U-Shaped Fire Ring

Smoke always seems to find a way to follow you, right? Here's a trick to help control it! Build your fire ring in a U-shape, with the open side facing where you want the smoke to go. Place larger rocks or a windscreen (like a stacked log or dirt mound) at the back of the U. Smoke tends to travel toward bigger objects, so this setup helps direct it away from where you're sitting while also reflecting heat back toward your campsite. Give it a try and enjoy a warmer, less smoky fire!

A Fire Needs Three Things to Burn:

1. **Heat** – A spark or ember to start it.

2. **Oxygen** – Air to keep it breathing.

3. **Fuel** – The food that keeps it burning.

So, we know that a fire needs a heat source, oxygen, and fuel to burn—but not all fuel is the same. And no, when we say fuel, we're not talking about gasoline or lighter fluid! A great fire isn't just about striking a match and hoping for the best. You need the right materials, in the right order, at the right time. Get this wrong, and your fire will struggle, smoke, or die out completely!

The Three Types of Fire Fuel

1. Tinder - The Fire Starter

Tinder is the stuff that catches the spark and starts the flame. It needs to be super dry and light so it burns easily—even from just a tiny spark.wa

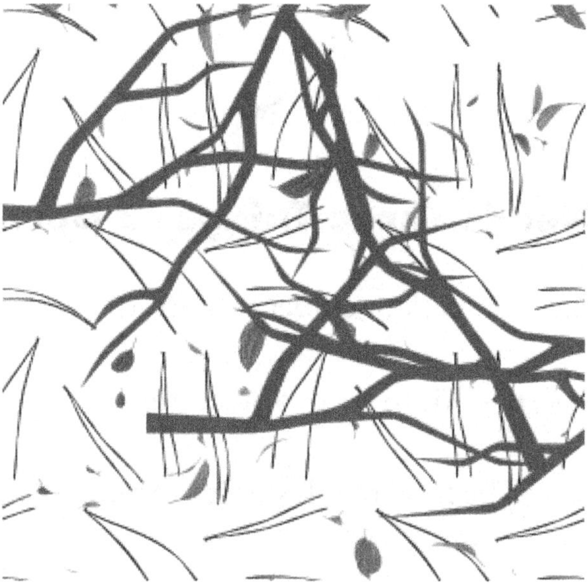

Best Tinder Materials:

- Fluffy plant fibers (cattail fluff, milkweed down)
- Dry leaves, pine needles, and tiny twigs (thinner than your pinkie!)
- Dry grass, lint, or bark shavings
- Cotton balls soaked in petroleum jelly (a survivalist's best trick!)

2. Kindling - The Fire Feeder

Once your tinder is burning, it needs help to grow. Kindling is small sticks and twigs that help the flames get stronger before adding big logs.

Best Kindling:

- Dry twigs and branches (thinner than your thumb)
- Small strips of bark (peeled off dead logs!)

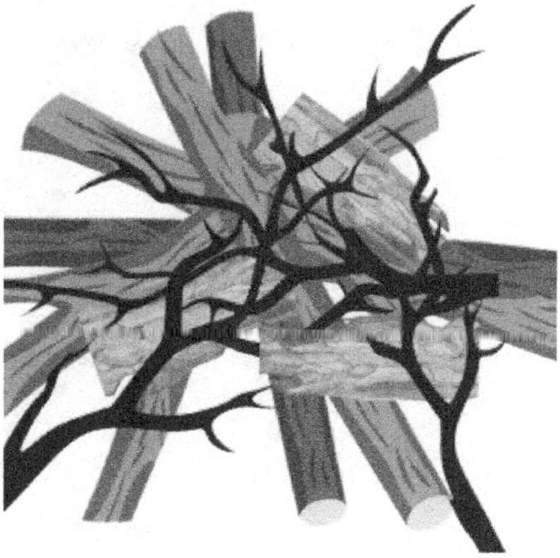

- Chopped-up branches that snap easily (bending = too wet!)

3. Firewood - The Fire Fuel

Once the fire is strong and steady, you can add larger logs to keep it burning for hours. These logs provide the steady heat and energy your fire needs to thrive.

Best Firewood Types:

- **Oak, ash, cherry, and maple** – Burn hot and steady (great for cooking!)
- **Hickory and birch** – Burn long and give off great heat

Avoid These Woods:

- **Pine, cedar, and fir** – Burn fast and pop loudly, sending sparks flying. Their sticky sap can also make your food taste bitter!
- **Green wood** – Freshly cut wood smokes a ton and takes forever to catch fire.
- **Standing dead trees** – These might look dry, but they could be home to animals like birds or squirrels. Use fallen wood instead!

Gathering Fuel for Your Fire

Before you even strike a spark, gather all your fuel—running out mid-way would be a disaster!

1. Two handfuls of tinder (about the size of a soccer ball).
2. A pile of kindling (at least the size of a basketball).
3. A knee-high stack of firewood to keep your fire going. The best firewood is as thick as your wrist—larger pieces take too long to burn.

*Pro Tip: Dry wood snaps—if it bends, it's too wet to burn! If you only have damp wood, place it near (but not too close to) the fire to help it dry. Just keep it far enough away to avoid stray embers!

Let the Sparks Fly—With Caution!

* **Fire Safety First!** Never start a fire or attempt to make sparks without an adult present. Fires can be unpredictable, and safety always comes first!

Now that you have the fuel your fire needs and plenty of air around you, what's missing? The ember!

An ember can come from another flame, like a lighter or match, or it can be created from a spark. A spark is a tiny, hot ember that needs some help to grow into a flame.

Sparks can be made in many ways. In the old days, flint and steel were commonly used. Nowadays, it's easier to use a magnesium fire starter. These tools can be tricky at first, so always practice safely with an adult.

Sparks need to be cared for because they aren't as strong as flames. First, they must be caught in a tinder nest, which shelters them and gives them fuel. Once your spark lands in the tinder, gently blow on it to help it grow into an ember and finally a flame.

* **Be careful!** Avoid breathing in smoke, and always have a trusted adult help you with this step.

How to Start a Fire Without Matches

Okay, so let's say you're in the wild, your matches got soaked, and your lighter ran out of fuel. Now what? No worries! People have been making fire without matches for thousands of years. Here's how you can, too!

Flint and Steel – The Ancient Spark Maker

What is flint and steel? Flint is a type of rock that, when struck against steel, creates sparks. These sparks are tiny pieces of hot metal that can ignite dry tinder and start a fire.

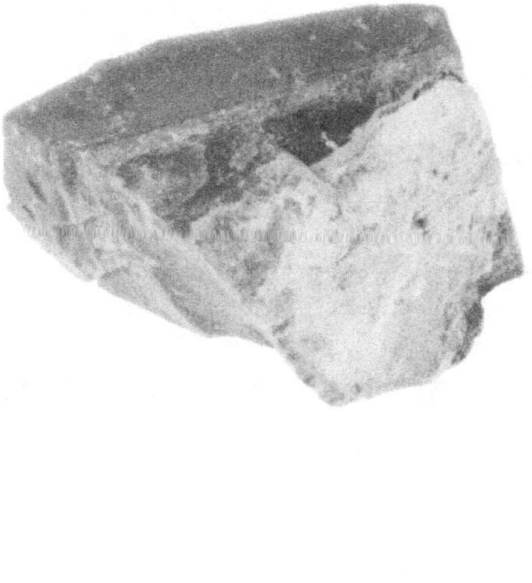

Where do you find flint?

- **In nature:** Flint is a hard, dark rock often found in gravel beds, riverbanks, or rocky areas. If you pick up a rock and strike it against another hard surface, and it makes sparks, that's your flint!
- **A Kit:** You can buy flint and steel fire-starting kits online or at outdoor stores.

How to Use Flint and Steel:

1. Gather dry tinder (shredded bark, dry grass, or cotton from your socks).
2. Hold the flint firmly in one hand and the steel striker in the other.
3. Strike the steel against the flint fast and hard, aiming the sparks at your tinder.
4. When a spark catches, gently blow on it to grow the ember into a flame.
5. Add kindling (small sticks) to feed the fire.

*** Pro Tip: If you can't find flint, try using the back of a knife blade against a hard rock!**

Magnesium Fire Starter – The Survivalist's Secret Weapon

What is a magnesium fire starter? It's a small metal tool that lets you make fire even in wet conditions.

It has two parts:

1. A magnesium block or rod (scrape off shavings to create fuel)

2. A steel striker (creates sparks when scraped)

Where do you find magnesium?

1. **Not in nature** – Unlike flint, you won't find a magnesium block lying around in the woods. If you don't have one, use other fire-starting methods.

2. **A Kit** – Magnesium fire starters are lightweight and found in most camping gear stores as a block or rod.

How to Use a Magnesium Fire Starter:

1. Scrape off a small pile of magnesium shavings onto your tinder.

2. Hold the steel striker at an angle and scrape fast to create sparks.

3. The sparks will ignite the magnesium shavings, creating a super hot flame.

4. Quickly add kindling and build up your fire.

*** Pro Tip: Magnesium burns at 5,610°F (super hot!), so be careful and ready to feed the fire fast!**

Fire by Friction – The Ultimate Survival Fire

If you don't have flint, steel, or a fire starter, you can still make fire by rubbing wood together! But be warned... this takes patience and muscle!

The Bow Drill Method - Best for beginners

This is the easiest friction fire method (but still hard). It uses a bow and spindle to spin wood fast enough to create an ember.

How to Make a Bow Drill Fire Starter:

What You Need:

- **A Fireboard** (a flat, dry piece of wood)
- **A Spindle** (a straight, dry stick about the length of your forearm)
- **A Bow** (a curved stick with a shoelace or paracord tied to it)
- **A Handhold** (a smooth rock or wood piece to press on top of the spindle to protect your hand)

How to Use a Bow Drill:

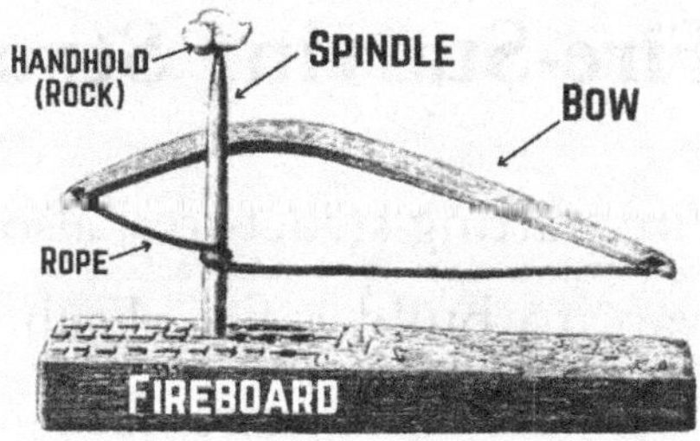

1. **Cut a small notch in your fireboard to catch embers.**
2. **Place a tiny pile of**

dry tinder next to the notch.

3. **Loop the spindle into the bowstring and place the tip in the notch.**
4. **Hold the handhold on top of the spindle to keep it steady.**
5. **Move the bow back and forth quickly to spin the spindle against the fireboard.**
6. **Keep going until you see smoke and a tiny ember in the notch!**
7. **Gently tip the ember onto your tinder, then blow on it until it catches fire.**

*** Pro Tip: The key is speed and pressure—move the bow fast but keep the spindle pressed down hard!**

Fire-Starting Structures

Fire-starting structures, also known as fire lays, are different ways to build a fire. Each structure serves a specific purpose—some are better for light, while others are ideal for cooking or long-lasting heat. Here are a few fire lays to get you started.

Teepee Structure

The teepee structure is great for providing light and allows the fire to grow upward. It's a reliable and easy-to-build fire, perfect for cooking food in a suspended pot.

To build a teepee fire:

- Make a nest of tinder (dry grass works best) and place it in the fire pit.
- Arrange a few pieces of kindling in a teepee shape around the tinder nest, leaving space between them.
- Add larger kindling around the smaller pieces.
- Place large firewood pieces over the structure, also in a teepee shape.
- Light the tinder nest inside the teepee with a match.

If the first layer of kindling doesn't catch fire well, have an adult carefully feed extra kindling or tinder to help it grow. Never place your hand close to the flames.

Lean-To Structure

Like a lean-to shelter protects from wind and rain, the lean-to fire structure shields a small flame while helping it grow. It's a great choice for cooking once the fire is burning steadily.

To build a lean-to fire:

- Lay a large piece of kindling on the ground inside the fire pit at a slight angle, with one end pointing into the wind.
- Place a tinder nest underneath the sloped kindling.
- Build a small teepee of sticks over the nest, keeping it under the sloped kindling.
- Add more kindling on either side to form a tunnel for airflow.
- Layer larger kindling over the structure, followed by firewood. Be careful not to let the heavy firewood collapse the kindling support.
- Light the tinder nest and monitor the fire, as wind direction can change.

Log Cabin Structure

The log cabin fire lay is easy to build and produces a long-lasting, high-heat fire. While it may take longer to catch, it provides consistent warmth. The more layers you add, the longer it will burn, but avoid making it too tall.

To build a log cabin fire:

- In the center of the fire pit, place a small teepee of kindling around a tinder nest.
- Lay two pieces of firewood on either side of the teepee.
- Stack two more pieces of firewood on top, running perpendicular to the first two.
- Continue layering, alternating directions, until you have five to eight layers.
- Gradually decrease the size of the firewood in the upper layers, finishing with kindling at the top.
- Keep small gaps between logs for airflow.
- Light the tinder through a gap at the bottom.

Platform Structure

When the ground is too wet, starting a fire can be tricky. The platform structure provides a dry base for your fire and produces steady, slow-burning coals, making it ideal for cooking. It's also called an "upside-down log cabin."

To build a platform fire:

- Lay three firewood pieces side by side with small gaps between them.
- Stack another three pieces on top, perpendicular to the first layer.
- Continue stacking four to five layers, then place two final layers with only two pieces of kindling.
- In the center of the top layers, place a tinder nest and surround it with smaller kindling in a teepee shape.
- Light the tinder and monitor the fire closely to ensure the structure catches.

Since this fire lay can be difficult to ignite, have an adult help add extra kindling if needed.

Star Structure

A star fire is a self-feeding fire that works best in a sloped fire pit. It burns steadily and is perfect for roasting marshmallows and telling stories around the campfire.

To build a star fire:

- Start by making a small teepee fire in the center of the fire pit.
- Arrange five long pieces of firewood around the teepee, pointing outward like a star.
- Light the tinder in the teepee and let it burn.
- As the fire grows, the heat and embers will ignite the surrounding logs.
- As the logs burn down, they will gradually feed into the center of the fire.
- If the logs burn away too quickly, ask an adult to add more firewood to keep it going.

Extinguishing a Campfire Safely

When the night ends, and it's time to head to bed, properly extinguishing your fire is essential. Simply throwing sand or water over the fire isn't enough—if done incorrectly, smoldering coals can reignite and cause a wildfire. Follow these steps to safely put out your fire:

- **Before starting a fire,** always have a bucket of water and a bucket of sand nearby.
- **When you're ready to extinguish the fire,** slowly pour water over the flames until they die out.
- **Stir the coals with a stick**—if they're still warm, they can reignite.
- **Pour sand over the damp coals** and mix thoroughly.
- **Look for red embers**—if you see any, scrape them off with an adult's help.
- **Hold your hand above the coals.** If you still feel heat, continue adding water and sand until the coals are completely cold.
- **Once the coals are cold,** the surrounding rocks should

also be cool to the touch.

- **Before leaving or going to bed,** scan the area to ensure no embers are smoldering.

Using both water and sand ensures the fire is completely out. Sand alone may smother coals temporarily, but they can stay hot for hours and reignite later, creating a wildfire risk. Always double-check before leaving a campsite.

Activity 1: Building and Lighting a Campfire

Fire is one of the most powerful survival tools you can master. In this activity, you'll put your skills to the test by gathering materials, constructing a fire, and lighting it—under the supervision of an adult, of course!

Before You Begin: Create Your Fire Checklist

Before you start building, make a checklist of the steps needed to build a safe fire. Here's a guide to help you:

Step 1: Choose a safe location.

Step 2: Clear the area of anything flammable.

Step 3: Find or dig a small fire pit and create a fire ring.

Step 4: Gather tinder, kindling, and fuel wood.

Step 5: Build a fire structure (teepee, log cabin, etc.).

Step 6: Have water or sand ready to extinguish the fire.

Step 7: Light the fire safely.

Use your checklist to stay on track! Think about each step carefully before moving on.

Put Your Knowledge to the Test! Ask yourself:

- Why is this location safe for a fire?
- What kind of fuel will help the fire burn best?
- Which fire structure works best for this situation?

Think carefully and make adjustments as needed. Your goal is to build a fire that lights easily, burns steadily, and can be controlled safely.

Ready to Light the Fire?

- Always have water or sand nearby to put it out if needed.
- Light your fire using your chosen method (match, lighter, or fire starter).
- Watch how the flames spread—what part of the fire catches first? How does the fire grow?

Activity 2: Homemade Fire Starters

Wilderness Warriors know that a fire won't start without dry tinder—but what if everything around you is damp? Instead of struggling to find dry materials, you can prepare your own supercharged fire starters that light easily and burn for a long time!

These fire starters are simple to make, store well, and will help you get a fire going even in tough conditions. Let's get crafting!

* Do not underestimate the messiness! Wax is HARD to clean if spilled!

What You'll Need:

- A bunch of dryer lint (it's flammable and makes great tinder!)
- Wax (old candles, crayons, or beeswax) A pressed paper egg carton *(No egg carton? No problem! Use a muffin tin with paper muffin liners instead.)*
- Wax paper or newspaper (to catch spills)
- A double boiler or a heat-safe bowl resting on a saucepan with water (for melting the wax)

- Scissors

* Safety First! This activity involves hot wax, so adult assistance is required. Wax can cause painful burns if spilled!

Step-by-Step Instructions

Step 1: Gather & Prep

- Collect lint from a dryer lint trap. Avoid clumping it too tightly—loose lint burns better!
- Cut or break wax into smaller pieces so it melts faster.
- Place small piles of lint into each cup of the egg carton.

Step 2: Melt the Wax

- With adult assistance, use a double boiler. If you don't have one, fill a pan with a few inches of water and place a heat-safe bowl or a snug-fitting pan on top.
- Add the wax to the top bowl or pan.

- Heat the water to a simmer, allowing the steam to gently melt the wax.
- Keep adding wax and stir as it melts until you have enough to pour.

Step 3: Pour the Wax

- Place the egg carton on wax paper or newspaper to catch any spills.
- Carefully pour the melted wax over the lint in each cup.
- If you don't have much wax, pour a little in each cup and repeat as needed. You want to soak the lint well so it burns longer.

Step 4: Let It Cool

- Leave the carton alone for about 45 minutes until the wax hardens.
- Once cool, cut out each individual cup—these are your fire starters!

How to Use Your Fire starters

These fire starters burn like mini candles—the wax keeps the lint burning for several minutes, giving your fire plenty of time to catch. However, keep in mind:

- You can't light them with a spark—you'll need a match or lighter.
- They work best when placed under your fire structure instead of dry tinder.

Activity 3: Roasting Over a Fire!

Roasting over a fire is an art—fun, relaxing, and totally worth the wait. Here are a few things to keep in mind as you cook over the flames:

1. **Be patient!** Great roasting happens near the coals, not in the fire or buried in the flames. Hold your food above the glowing embers, not where it will scorch instantly.
2. **Oops! Dropped your food?** That's not ideal, but it's not the end of the world. Make it an experiment—watch and see how long it takes for the fire to consume it.
3. **Fire on your food?** Don't wave it around! That will only send flaming marshmallows or sausages flying into

someone's lap. Instead, blow it out gently or smother it against a non-burnable surface (like a rock).

4. **Enjoy the experience!** Campfire cooking is all about the adventure—even when things don't go as planned!

Choose Your Adventure:

Adventure #1: Classic S'mores

A gooey, chocolatey favorite!

You'll Need:

- Marshmallows (store-bought or homemade!)
- Graham crackers (snapped in half)
- Chocolate bars (same size as the cracker halves)
- Fire-safe skewers or roasting sticks

How to Make It:

1. Put a marshmallow on your skewer.
2. Hold it over the fire, about six inches above the glowing coals. Slowly rotate until golden brown (or charred if you like it extra toasty!).

3. Place the roasted marshmallow on a piece of chocolate sitting on a graham cracker.

4. Sandwich it with another graham cracker and give it a gentle squeeze.

5. Let it cool slightly, then enjoy!

* *Want to mix things up? Try different flavored chocolates or use cookies instead of graham crackers!*

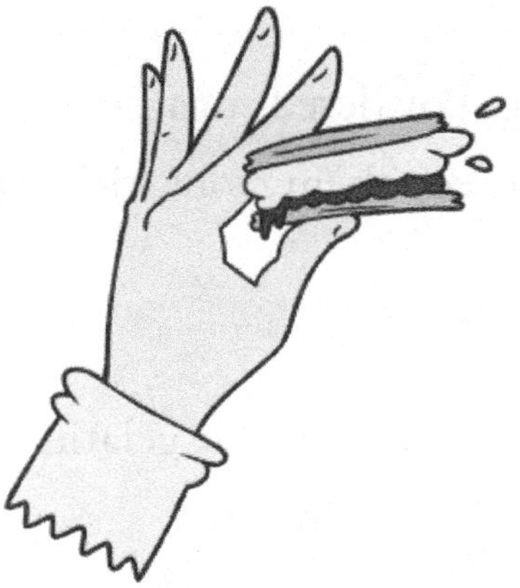

Adventure #2: Homemade Marshmallows (Fluffy and fun to make!)

Store-bought marshmallows work fine, but homemade ones take your s'mores (and roasting skills) to the next level!

You'll Need:

- 3 tablespoons (or 3 packets) unflavored gelatin
- 1 cup honey or maple syrup
- ½ cup water
- 1 teaspoon vanilla extract
- ¼ teaspoon salt
- For dusting: Powdered sugar, cornstarch, or arrowroot powder (to keep them from sticking)

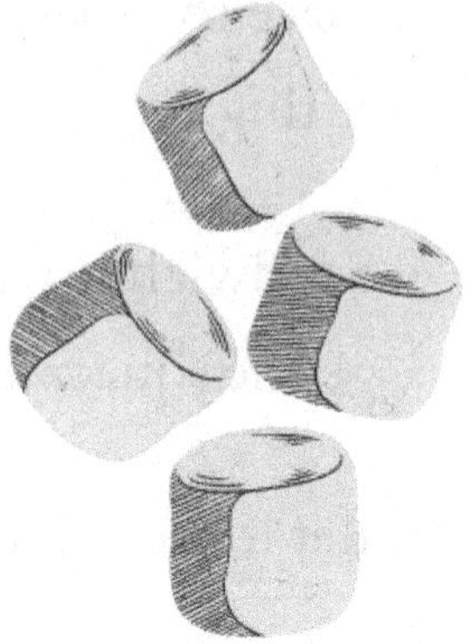

How to Make It:

1. In a mixing bowl, sprinkle gelatin over ½ cup of water and let it sit for 5 minutes.
2. In a saucepan, heat honey or maple syrup until it reaches 240°F (use a candy thermometer).
3. Slowly pour the hot syrup into the gelatin mixture while mixing on low speed. Then, increase to high speed and whip for about 10 minutes, until it looks thick and fluffy.

4. Stir in vanilla extract and salt.

5. Grease a pan and dust it with powdered sugar, cornstarch, or arrowroot. Pour the marshmallow mixture in and smooth it out.

6. Let it set for at least 4 hours before cutting into squares. Dust the pieces with more powdered sugar, cornstarch, or arrowroot to prevent sticking.

*** Pro Tip: You can eat them right away, but for the best roasting marshmallows, let them sit out uncovered for a few days after cutting. This helps them dry out a bit, making them toast up golden and gooey instead of just melting off the stick!**

Adventure #3: Campfire Sausage Roast (A great non-sweet option!)

For those who prefer a savory snack, nothing beats a perfectly roasted sausage over the fire.

You'll Need:

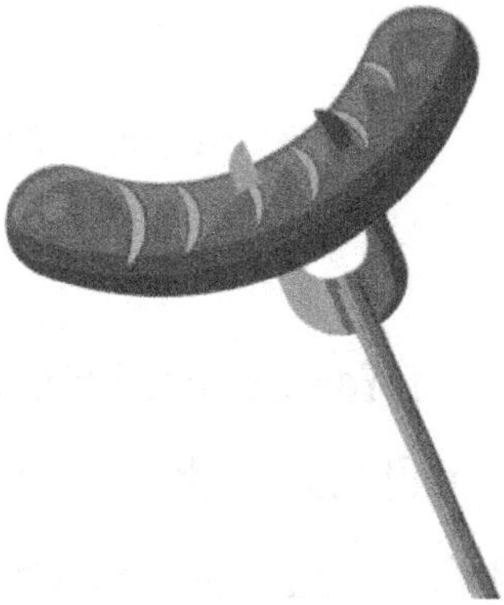

- Sausages (your favorite kind!)
- Fire-safe skewers or roasting sticks
- Mustard, ketchup, or any favorite toppings

How to Make It:

1. Push the sausage onto your skewer.
2. Hold it over the fire, about 6 inches above the coals. Rotate slowly for even cooking.
3. Roast until it's browned and cooked through (about 5–10 minutes).
4. Let it cool slightly, then enjoy with your favorite toppings!

Wilderness Warrior Skill Unlocked!

You've mastered the art of fire-Making! Now, mark your progress by coloring in your Wilderness Warrior Fire Master award on your Survivalist Skill Tracker (page 277).

Level Up Your Survival Skills!

Now you know how to build a fire and stay warm in the wild! But fire alone won't keep you alive—you NEED water! Knowing where to find clean drinking water is a game-changer.

Up next: Skill 4 – Finding Water Like a Survival Expert! Let's GO!

Survival Skill 4: *Finding Water*

Only about 3% of the world's water is freshwater, and less than 1% is easily accessible for drinking. The rest is locked away in glaciers, ice caps, or deep underground. That's why learning to find clean water is one of the most essential survival skills.

Water: The Lifeline of Survival

You've got fire. You've got shelter. But without water, you won't last long.

Water is life—literally. Your body is made up of about two-thirds water, and every system inside you depends on it to function. Remember the Rule of Three? You can survive weeks without food, but only about three days without water before dehydration takes its toll. Push beyond that, and things get dangerous fast.

Even if you're not lost in the wild, dehydration can start sneaking up on you in as little as six hours without a drink. You might feel thirsty at first, but soon, your mouth dries up, your energy drains, and headaches start pounding. The more active you are—running, climbing, or even just breathing in dry air—the faster you lose water. Even wind and cold temperatures steal moisture from your body.

So, what do you do? You find water. But not just any water—clean water. Drinking from a random lake, river, or puddle might seem like a good idea when you're desperate, but invisible bacteria and parasites can turn a sip into a survival nightmare. That's why learning to find, collect, and purify water is one of the most important wilderness skills you'll ever master.

Let's dive in!

Hunting for Hidden Water

Think finding water is easy? Think again! Not all water is sitting out in the open like a sparkling lake or a rushing river. Sometimes, you have to play detective and follow the clues nature leaves behind.

Clues That Lead to Water

The best way to track down water is by looking for signs of wet ground or water-loving plants like cattails. If you see those, there's a good chance water is nearby. But what if the ground looks dry? No worries—there are other clues to follow!

- **Animal Footprints:** Many animals head to water in the early morning. If you find fresh tracks leading in one direction, you might be on the right path! Just be careful—tracks can also lead to food or shelter.

- **Bird Behavior:** Seed-eating birds, like pigeons or sparrows, often travel toward water sources during dry seasons. If you see a lot of birds flying in the same direction, they might be leading you to water.

- **Insect Swarms:** Some bugs, like midges or mosquitoes, love to hang around water. If you notice a cloud of tiny insects buzzing in one spot, there might be water nearby!

While these signs aren't foolproof, they can help point you in the right direction. But sometimes, you need more than clues—you need a solid plan to track down that life-saving water!

Where to Look for Water

Above-Ground Water

The easiest kind of water to find is above-ground water, like lakes, rivers, and streams. But don't just scoop it up and start chugging—you don't know if it's safe! Fast-moving water is usually cleaner than still water, but even rushing streams can have invisible dangers like bacteria or parasites. If you find water, collect it and purify it before drinking.

Morning Dew

Did you know you can drink the morning dew? Yep! Just wrap a cotton shirt around your arms and legs and take a walk through tall grass early in the morning. This is called the **sweep-and-collect method**—as you walk, your clothes absorb the dew. Then, just squeeze the water into a container. Easy, right?

Snow and Ice

If you're in a freezing environment, you might think, "Easy! I'll just eat snow!" Bad idea. Eating snow lowers your body temperature, making you even colder—definitely not what you want in a survival situation. Instead, melt it first. The best trick? If you have ice, melt that in a pot over a fire first, then add snow. This stops the water from tasting burnt!

What about sea ice? Normally, drinking salt water is a big no-go. But did you know that old sea ice—the kind that looks blue—has much less salt? If you find some, you can melt it down for drinking water! Just make sure it's the right kind, or you'll end up thirstier than before.

Rainwater Collection

Rainwater is one of the safest sources of drinking water—if you can catch it! If you have plastic or a tarp, tie it between some trees and place a rock in the center. Rain will run down the tarp and pool around the rock, making it easy to collect. No tarp? Big leaves work too—just use them to funnel water into a container.

Finding Underground Water

Sometimes, water is right under your feet, but you just can't see it! Here's how to dig for it:

1. Look for wet soil in low-lying areas or dips in the land.

2. Spot plants like cattails—they only grow where their roots can reach water.

3. Start digging! Use a sturdy stick or spade and watch for wet sand or mud. If water starts trickling down the sides of the hole, you're on the right track!

4. Let the hole fill up naturally. Dig a few more sip wells while you wait.

Warning: This water will probably be muddy, so you'll need to filter and purify it before drinking.

Juice of the Wilderness – Plants with Drinkable Water

Some plants store water or juice that you can drink! But be careful—not everything is safe.

- **Prickly Pear Cactus** – The fruit (prickly pear) can be squeezed for juice. Just watch out for the spines!

- **Coconuts** – Green coconuts are packed with two cups

of coconut water—a great drink! But mature brown coconuts have thicker milk, and you shouldn't drink more than three a day.

- **Fruits and Berries** – Some fruits and berries have a lot of water, but only eat them if you 100% know they are safe. Eating too many can upset your stomach!

- **Maple and Birch Trees (especially Paper Birch and Yellow Birch)** – In early spring, you can tap (make a hole in) maple and birch trees to collect fresh, slightly sweet sap. It's full of nutrients and safe to drink!

- **Willow Trees** – While willow bark is better known for medicinal uses, their roots often indicate nearby water sources. Some species store small amounts of drinkable moisture in their branches.

- **Bamboo** – In tropical and jungle areas, some species of bamboo store fresh water inside their hollow segments. Look for young, green bamboo and carefully cut a section to access the water inside.

*** Pro Tip: Need to open a coconut but you don't have any tools? No problem! Find a big, sharp rock or a hard, uneven surface. Hold the coconut with both hands and slam it against the edge, rotating it**

slightly after each hit. If you're on higher ground, dropping it onto a sharp rock can help. Once it cracks, pry it open with your hands or a sturdy stick.

Water Sources to Avoid (Unless Desperate)

Before you take a sip, use your **eyes, nose, and brain** to decide if the water is safe:

- **Does it smell fresh?** If it stinks, skip it!
- **Are animals and insects around?** No fish, birds, or bugs? That's a red flag.
- **Is the water moving?** Still, stagnant water is a breeding ground for bacteria.

Some water sources are risky even if they look clean. Watch out for signs of pollution like oil slicks, farm pesticides, or trash floating nearby. Even if the water looks crystal clear, chemicals and waste can still make it dangerous to drink. Here's what to avoid unless it's life or death:

- **Stagnant Water** – Ponds, puddles, and slow-moving water often hide bacteria and parasites.

- **Too-Peaceful Water** – If no birds, bugs, or animals are drinking from it, that could mean contamination.
- **Water Near Animal Tracks** – A good sign animals are drinking, but also a high risk for bacteria from waste. Purify it!
- **Bright Green or Scummy Water** – Algae-covered, foamy, or oddly colored water? That's toxic.
- **Snow/Ice** – Raw it'll lower your body temperature—always melt it before drinking.
- **Ocean Water** – Saltwater is a trap—it dehydrates you instead of quenching your thirst.

Pro Tip: Crystal-clear water isn't always safe! Bacteria, viruses, and parasites are invisible. Always purify your water before drinking.

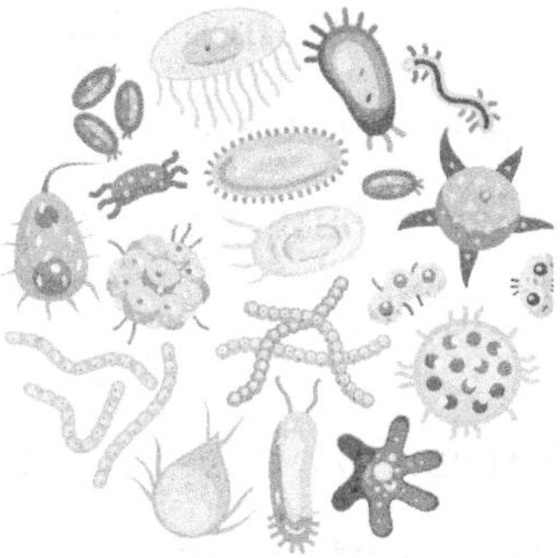

How to Make a Water Filter

Clean water is important for survival! Unless you're lucky enough to find a crystal-clear mountain spring, you'll need to filter and purify your water before drinking it.

Step 1: Filtering Water – Removing the Gunk

Filtering removes dirt, bugs, and debris, making the water clearer—but it won't remove bacteria or viruses. That's why purifying comes next!

What You'll Need:

- A plastic bottle (or a hollow log with the bottom cut off)
- A sharp knife or scissors (ask an adult for help)
- A piece of cloth (bandana, clean shirt, or even grass if no cloth is available)
- Charcoal (from a fire) – Helps remove toxins
- Sand (fine or coarse—whatever you can find) – Filters out dirt

- Small rocks or gravel – Catches larger debris
- Dirty water (for testing!)
- Two containers (one for pouring, one for catching clean water)

How to Build Your Filter:

Prepare the Bottle:

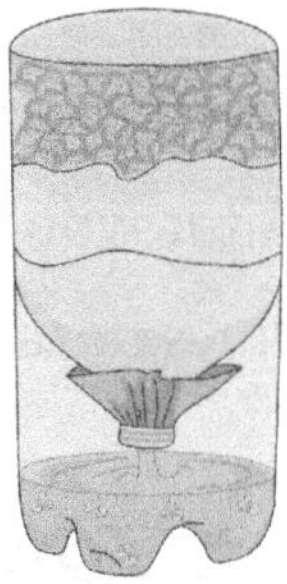

- If using a plastic bottle, cut it in half.
- Flip the top half upside down and place it inside the bottom half. (The top half will be your filter, and the bottom half will catch the clean water.)
- If your bottle has a pull-top cap, leave it closed or poke a tiny hole for water to drip out slowly.
- No cap? Cover the neck with cloth and secure it with string or a rubber band.

No Bottle? No Problem!

- Find a hollow log, sturdy bark, or large leaf.

- Shape it like a cone or tube and layer your filter materials carefully inside.

Layer Your Materials in This Order:

- Bottom layer (near the opening): Cloth or grass → Catches tiny particles.
- Next layer: Crushed charcoal from a fire → Absorbs toxins and chemicals.
- Next layer: Sand → Filters out fine dirt.
- Top layer: Small rocks or gravel → Catches big debris like leaves and bugs.

Pour the Dirty Water:

- Slowly pour the dirty water into the top of your filter.
- Let it drip through all the layers into your clean container.

Repeat If Needed:

If the water still looks dirty, pour it through the filter again.

Step 2: Purifying Water – The Final Step

Even filtered water can have invisible dangers like bacteria and viruses. Here's how to make sure your water is truly safe:

1. Boiling (The Ultimate Power Move!)

Boiling is the best and easiest way to purify water. Just follow these steps:

- Pour your filtered water into a pot.
- Heat it over a fire until it reaches a rolling boil (big, constant bubbles).
- Let it boil for 5 minutes—this kills bacteria, viruses, and parasites.
- Let it cool before drinking.

How to Boil Water Without a Rack

If you need to boil water but don't have a metal rack, try these hot survival tricks!

Rock Boiling (Best If You Don't Have a Fireproof Container!)

This is one of the coolest survival tricks—and you'll love watching the water heat up with just hot rocks!

- Find dry, non-porous rocks (avoid river rocks, which can explode).
- Heat the rocks in a fire for about 30 minutes until very hot.
- Carefully transfer the rocks into your water using sticks or tongs.
- Watch the magic! Keep adding hot rocks, and soon your water will boil!

Primitive Method (For When You Have a Metal Container)

This is the easiest way to boil water if you have a metal cup or pot—just put it straight into the fire!

- Let the fire burn down to hot embers (coals work better than flames).
- Set your metal container in the embers and wait for the

water to boil.

- Use sticks or gloves to carefully remove the pot once it's ready!

Dakota Fire Hole (Best for Windy Days!)

This underground fire pit keeps flames protected from wind while still being hot enough to boil water.

- Dig a hole about 12 inches deep and wide.
- Dig a second hole about a foot away, connecting it to the first hole underground (this creates a tunnel for airflow).
- Build a fire in the main hole and let it burn down to hot coals.
- Place a flat rock or metal pot over the hole and let the fire do its work!

2. Purification Tablets (Tiny but Mighty!)

If you have water purification tablets, follow the instructions on the package. These are great for emergencies!

Pro Tip: There are nifty gadgets like the LifeStraw that filter out bacteria and dirt, making water safe to drink instantly. If you're heading into the wild, keeping one in your backpack could be a lifesaver!

Distilling Water: Turning Saltwater into Fresh Water

If you're ever stranded near the ocean, drinking seawater is a huge mistake (it'll make you even thirstier!). But don't worry—you can turn saltwater into fresh water using heat and condensation.

What You Need:

- A pot or container of seawater
- A fire (or hot sun if you're patient)

- A raincoat, tarp, or a large, waxy leaf (to trap steam)
- A smaller container to collect fresh water

How to Do It:

1. Heat the seawater over a fire until it starts steaming.

2. Trap the steam using your raincoat, tarp, or a big leaf. (Hold it above the pot but don't let it touch the fire!)
3. Tilt it downward so the water droplets run toward your collection container.
4. Watch the science happen before your eyes! The steam turns back into fresh water, leaving the salt behind.

This method works because pure water turns into steam, but salt doesn't! The result? Fresh, drinkable water.

Pro Tip: If you don't have fire, you can let the sun do the work by placing seawater in a shallow hole, covering it with plastic or a raincoat, and putting a small rock in the center so condensation drips into a cup. It's slow, but it works!

Activity 1: Find Water in the Wild

Out in the wilderness, water can hide in unexpected places. The big question is—if you were lost, could you find enough to survive?

Your Mission:

Use survival tracking techniques to locate a natural water source and collect some water. Some will be obvious—others, not so much!

Here's how:

- **Scout the Land** – Using the skills we learned earlier, search for signs that water might be nearby.
- **Safety Tip!** Not all water is safe—even if it looks clean. Avoid collecting it near roads, farms, or anywhere chemicals might run off. If you're not sure, skip it.
- **Leave a Trail** – Practice marking your path using

sticks, stones, or other natural markers so you don't get lost (just like Hansel & Gretel—but smarter).

- **Dig for Water** – If you find damp ground, dig a small hole and wait. Does water start to collect? Congratulations! You've found a hidden water source!

- **Collect Your Water** – Gather your water in a cup, jar, or container (you'll need it for the next activity).

- **Analyze It** – Does it look clear or dirty? Does it smell weird? (Reminder: Never drink it until it's been filtered and purified!)

Let's find out how clean we can make it in the next challenge!

Pro Tip: If the ground is moist but no water appears, line your hole with leaves and check back later—water may slowly seep in!

Activity 2: Science Challenge: Build a Water Filter

Now that you've found a water source, let's clean it up!

You're going to make your own water filter and test how well it works. This helps remove dirt and debris—though don't forget, filtering alone won't make it safe to drink. That's what we'll tackle in the next activity!

What You'll Need:

- A water bottle, funnel, or a clean cup with a small hole in the bottom
- A coffee filter or clean cloth
- Sand, small rocks, or charcoal (optional for advanced filtering)
- Dirty water
- Three clear cups or jars (One for dirty water, one to catch the filtered water, and one for comparing a second attempt or clean tap water.)

Step 1: Build Your Filter

You can build a filter just like the one described earlier in this chapter—using a plastic bottle, cloth, charcoal, sand, and rocks.

Or, make a quick version by layering your materials into a funnel or paper cup:

Layer in this order (bottom to top):

1. Coffee filter or paper towel – catches fine particles

2. Crushed charcoal – absorbs toxins

3. Sand – filters out dirt

4. Small rocks or gravel – traps leaves and bugs

Step 2: Test It Out!

Pour your dirty water into the top of the filter and watch it drip into the clean cup below.

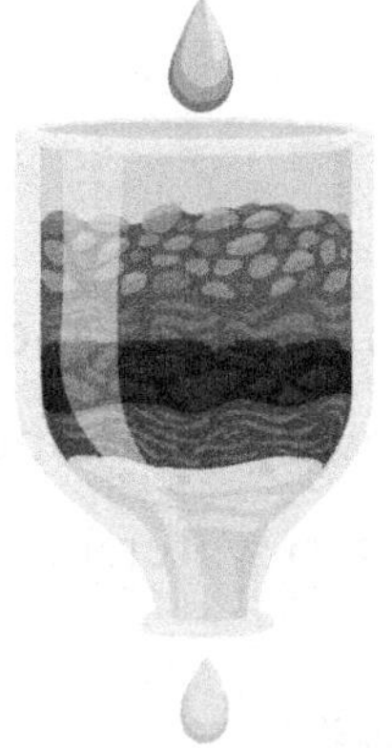

Be patient—it might drip slowly, but that means it's working!

Step 3: Compare Results

Hold your clean and dirty cups side by side.

Observe carefully:

- Does the filtered water look clearer?
- Is the dirt gone?
- Does it smell better?

Try pouring the filtered water through again to see if it gets even cleaner.

Looks better, right? But remember, germs are invisible to our eyes. Let's take it to the next level in our final challenge—boiling!

Activity 3: Fire & Water: Purify Like a Pro

If you were really lost in the wilderness, you wouldn't just want clean-looking water—you'd want safe water. Filtering isn't enough.

Your final mission: Purify it!

With an adult's help (because burns and bacteria are no fun!), it's time to boil your water to safely remove bacteria and germs.

Here's how:

1. Boil your filtered water in a pan for at least 5 minutes to kill bacteria and germs.
2. Let it cool.
3. Now you've successfully filtered and purified your water—awesome!

Observe and ask yourself:

- Does it look clearer?
- Smell better?
- Would you feel confident drinking it if you were in a real survival situation?

If yes—awesome! You've successfully filtered and purified your water like a true Wilderness Warrior.

Important Note:

This is how we would do it in a survival situation. Since we're not actually in the wild right now—and because boiling does not remove certain dangerous things like chemicals, oils, tannins, or minerals—we will not drink it for safety reasons.

Instead, celebrate your success by treating yourself to clean water from your home faucet. Isn't it awesome to have easy access to safe water?

Want more hands-on fun?

Try making your own **Hot Cocoa** from scratch using the recipe on page 318!

Try This With a Group! – Flash Flood Survival Challenge!

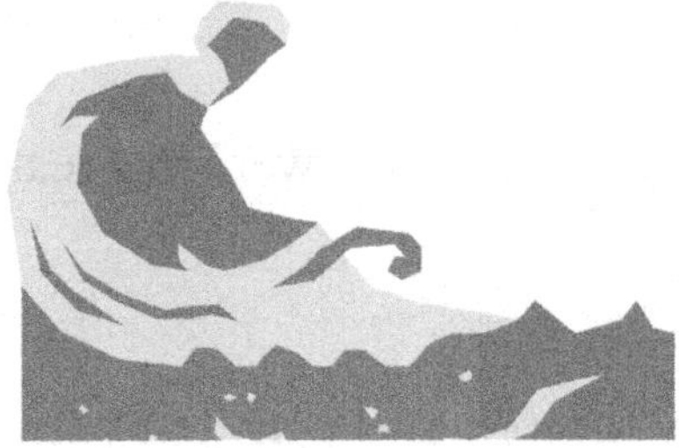

Oh no! A flash flood is coming! You have 10 seconds to find higher ground before the water sweeps you away!

How to Play:

1. Choose one person to be the Flash Flood Caller.
2. When they yell "FLASH FLOOD!" everyone has 10 seconds to climb onto a rock, fallen log, or other high ground.
3. If someone doesn't make it in time, they've been caught by the flood! The caller helps them get to safety.
4. Keep playing, changing locations for new survival challenges!

Where would you go if this were real? What makes a spot safe?

Wilderness Warrior Skill Unlocked!

You've mastered the art of finding, filtering, and purifying water! Now, mark your progress by coloring in your Wilderness Warrior Water Master Award on your Survivalist Skill Tracker (page 278).

Level Up Your Survival Skills!

You've got fire, shelter, and water—awesome! But what if a storm is rolling in? Now you'll learn to read the sky, predict wild weather, and stay one step ahead of nature!

Up next: Skill 5 – Mastering the Weather! Let's GO!

Survival Skill 5: Mastering the Weather

"Look deep into nature, and then you will understand everything better." — Albert Einstein

Weather: Predicting the Unpredictable

Imagine you're out on an adventure—maybe hiking through the woods, camping by a lake, or exploring a new trail. The sun is shining, birds are singing, and everything seems perfect. But suddenly, dark clouds roll in, the wind picks up, and before you know it—BOOM! A thunderstorm crashes overhead. If you weren't prepared, your trip just took a soggy turn for the worse.

That's why knowing how to predict the weather is a skill every Wilderness Warrior should have! Even if you check the forecast before heading out, the weather can change in an instant. Fortunately, the world around you is full of clues that can help you figure out what's coming next—you just have to learn how to read them!

How to Predict the Weather – Clues Are Hidden in Plain Sight!

Clues about the weather are all around you—you just have to pay attention! These quirky tricks aren't perfect, but with practice, you might out-guess a meteorologist!

- **Cricket Thermometers** – Want to know the temperature without a phone? Crickets can tell you! Count how many times a cricket chirps in 14 seconds, then add 40. That's the temperature in Fahrenheit. Try it a few times and average your results for the most accurate reading.

- **Spiders and Ladybugs on the Move** – Spiders hate the cold! If you start seeing more of them creeping indoors, that could mean colder weather is on the way. Ladybugs feel the same way, but instead of coming inside, they huddle together in big groups to stay warm. Spot a loveliness (yes, that's the real name for a group of ladybugs), and you might need an extra layer soon!

- **Buggy Behavior Before a Storm** – Ever noticed mosquitoes and black flies attacking in full force, then suddenly disappearing? That's because these little pests can sense storms hours before they arrive. If bugs are extra annoying, bad weather could be coming within the next 12 hours. If they suddenly vanish—get ready! The storm is about to hit.

- **Pinecones** – The Forest's Weather Report – Pinecones are like tiny weather forecasters. When the air is dry, they open up to release their seeds. But when rain is coming, they slam shut like little umbrellas to protect their seeds from getting soaked.

- **Lightning Countdown** – When you see a flash of lightning, start counting the seconds until you hear thunder. Take that number and divide it by 5. The answer tells you how many miles away the lightning strike was. If the time gets shorter with each flash, the storm is getting closer—time to take cover!

- **Birds Know Best** – Birds don't like flying in storms,

and they know when bad weather is coming. If you notice birds flying lower than usual or staying close to the ground, it's a sign of low air pressure, which means a storm might be on the way.

- **Campfire Smoke Clues** – If the smoke from your campfire rises straight up, the weather is likely to stay clear. But if the smoke curls downward and lingers near the ground, low pressure is moving in, which often means rain is coming.

- **Dew in the Morning** – If you wake up early and see heavy dew on the grass, you're in for a nice, clear day. If there's no dew at all, rain could be on the way.

One of the best ways to learn about the weather is to talk to people who have lived in the area for a long time. They know all the local weather patterns and might have some cool tricks they've learned over the years. So next time you're around a seasoned outdoorsman, ask them about their favorite weather prediction tips!

Types of Clouds and What They Mean

High-Level Clouds (20,000 feet and up!)

These clouds live way up high, where the air is thin and chilly.

Cirrus – Wispy, feathery clouds made of ice crystals. If the sky looks like someone dragged a paintbrush across it, you're looking at cirrus clouds. They often mean a change in the weather is coming in a day or two!

Cirrostratus – Thin, white clouds that stretch across the sky like a sheet. They usually show up 12–24 hours before rain or snow. Keep an eye out!

Cirrocumulus – Tiny, puffy clouds arranged in rows, like fish scales. In cold weather, they can mean a storm is on the way. In the tropics, they can be a warning sign of a hurricane!

Middle-Level Clouds (6,500–20,000 feet up!)

These clouds are lower than cirrus clouds but still high in the sky.

Altostratus – Blue-gray clouds that blanket the sky. If they show up, a storm is likely on its way.

Altocumulus – Large, puffy clouds that look like cotton balls. If you see them in the morning on a warm, humid day, you might get an afternoon thunderstorm!

Low-Level Clouds (Up to 6,500 feet—really close!)

These are the clouds you can see best—and the ones most likely to bring rain.

Stratocumulus – Big, puffy clouds that form in rows with blue sky in between. They don't always bring rain, but they can turn into storm clouds.

Cumulus – These are the classic, cotton-candy clouds. If they stay small, the weather will likely stay nice. But if they grow tall, get ready for a storm!

Nimbostratus – Huge, dark clouds that cover the sky. If you see these, grab a raincoat—rain or snow is on the way!

Storm Clouds & Oddballs

Some clouds are wildcards—you definitely want to watch out for these!

Cumulonimbus – These towering clouds mean business. They can stretch from the ground all the way to the highest levels of the sky, bringing thunderstorms, lightning, hail, and even tornadoes. Watch the top—if it looks like an anvil, that's the direction the storm is moving!

Mammatus – Bubble-shaped clouds hanging from the bottom of a storm cloud. If you see these, get ready—bad weather is on the way.

Lenticular – These UFO-shaped clouds form over mountains and hills. No, they're not aliens, but they sure look cool!

Contrails – These thin white streaks are left behind by airplanes. If they disappear quickly, the air is dry and clear. If they stick around, the air is moist—rain might be coming!

Fractus – Shredded-looking clouds that appear like torn pieces of fluff. These mean the wind up high is super strong!

Green Clouds – If the sky turns green, take cover! These rare clouds mean a severe storm, hail, or even a tornado is possible.

Not All Clouds Mean Trouble!

Some clouds are completely harmless. Puffy cumulus clouds often mean the weather is great, and even wispy cirrus clouds just tell you change is coming—not always a storm. The more you watch the skies, the more you'll learn to tell the difference!

Cloud Name Decoder – Can You Crack the Code?

Cloud names sound fancy, but they're really just made of simple clues. Here are a few to remember:

Cirrus = high

Alto = middle

Cumulus = heap or pile (usually low)

Stratus = layer (like a blanket)

Nimbus = rain or storm cloud

Now let's test your skills!

What does *stratocumulus* mean? You guessed it—a low, lumpy blanket of clouds in the sky!

What about *altostratus*? That's right—a mid-level, spread-out layer of clouds.

Try one more: What might *nimbostratus* be? Yep—a rain cloud that spreads across the whole sky like a soggy blanket!

Preparing for Weather Changes

Weather changes aren't just inconvenient—they can be downright dangerous if you're not prepared! Imagine getting ready for a fun, sunny day, only to step outside and find a snow-covered campsite. If you're not dressed for the cold, you'll be miserable—and possibly forced to cut your adventure short.

That's why every Wilderness Warrior should be ready for sudden weather shifts. The good news? A little preparation goes a long way!

Check the Forecast Before You Go

Before packing for any outdoor adventure, check the weather forecast. Apps like the National Weather Service (NWS) can give you a good idea of what to expect in the days ahead. Look at:

- **Temperature** – This helps you decide what clothes to bring for both day and night.
- **Weather Warnings** – Pay attention to alerts for storms, extreme temperatures, or other hazards.

- **Elevation Changes** – The higher you go, the cooler and wetter it tends to be.

Wind Speed and Direction Matter

Wind can affect more than just how your hair looks! Strong winds can make campfires unsafe and impact activities like windsurfing or kayaking. Knowing the wind direction can also help you plan ahead.

Pack Smart, Stay Safe

Even with the best forecast, weather can change unexpectedly. That's why a weather emergency kit is a smart idea—you can even keep one in your car! Here are some essentials to include:

- Emergency blanket
- First aid kit
- Headlamp or flashlight (with extra batteries)
- Sunscreen
- Extra clothing for sudden temperature drops

Always Have Extra Food & Water

Getting stranded in bad weather is no fun, but it's even worse if you're hungry or thirsty. Pack nonperishable (long shelf life) snacks and plenty of water so you're always prepared.

By planning ahead, you won't just stay safe—you'll be ready for whatever comes your way!

* Surviving Extreme Weather *

Okay, now you know how to **predict** the weather and how to prepare for weather changes. But what if you get **stuck in it**? Here's what to do in dangerous conditions:

Thunderstorms – When Lightning Strikes!

A thunderstorm isn't just a rainstorm—it comes with thunder and lightning, which makes it much more dangerous. While rain and wind can cause trouble, lightning is the real danger in a thunderstorm.

But don't worry! By knowing what to do, you can stay safe whether you're at home or outdoors.

How to Stay Safe in a Thunderstorm:

- **Get indoors as fast as possible.** If a building is nearby, head inside immediately.
- **If no shelter is available, get low.** Lightning strikes the highest point, so move downhill or find a lower spot in the land. But **avoid dips** that could flood.
- **Stay away from water.** Water conducts electricity, so keep at least 330 feet (100 meters) away from lakes, rivers, or puddles.
- **If you're near a car, get inside.** Cars aren't as safe as buildings, but they can protect you—just **don't touch the metal** parts!

Where Not to Take Cover:

- Under a lone tree—lightning is more likely to strike the tallest object.
- In a tent or pavilion—many have metal parts that conduct electricity.
- Near metal objects—metal doesn't attract lightning, but it spreads the electricity if struck.

Signs That Lightning is About to Strike

- A strange crackling or buzzing sound in the air
- Your hair standing on end (like when you rub a balloon on it)
- A tingling sensation on your skin

If you notice any of these signs, act fast!

The Lightning Safety Position

If you're caught in the open with no shelter:

1. **Place something (like your backpack) between you and the ground.** *Lightning can travel through the ground after a strike. Insulating yourself with a dry, non-metallic object helps reduce the amount of electricity that reaches your body.*

2. **Squat low, keeping only the balls of your feet on the ground, with your heels touching.** *Staying low makes you a smaller target, and minimizing contact with the ground reduces the risk of ground current passing*

through your body.

3. **Tuck your head down to your knees, wrap your arms around your legs, and cover your ears.** *This position helps protect your vital organs and reduces exposure to loud thunderclaps, which can cause hearing damage.*

4. **Close your eyes and hold your breath as long as possible.** *If lightning strikes nearby, the bright flash can temporarily blind you, and the sudden pressure change can affect your breathing. Closing your eyes and holding your breath helps minimize these effects.*

By staying aware and knowing what to do, you can keep yourself safe even in the most unexpected storms!

Pro Tip: If you're in a group, spread out at least 100 feet apart so lightning doesn't jump between people.

Tornadoes – When the Sky Starts Spinning!

Tornadoes can form quickly during severe storms, creating powerful, twisting winds that can lift cars, tear off roofs, and toss heavy objects like toys. If the sky turns greenish, the wind suddenly stops, or you hear a loud roar like a freight train—take cover immediately!

How to Stay Safe During a Tornado:

- **Get underground if possible.** A basement or storm shelter is the safest place to be since tornadoes can rip buildings apart.

- **No shelter? Find the lowest ground.** Lie flat in a ditch or low-lying area, covering your head and neck with your arms to protect from flying debris.

- **Stay away from cars and trees.** Tornadoes can lift heavy objects, and trees can snap like twigs in the high winds.

- **Avoid windows and wide-open spaces.** If you're indoors without a basement, move to a small, windowless room on the lowest floor (like a bathroom or closet).

Pro Tip: If you see debris swirling in the distance, a tornado may be forming—don't wait! Get to safety immediately.

Fighting the Freeze – How to Stay Warm When It's Freezing!

When the temperature drops, the cold can turn dangerous fast. But with the right tricks, you can beat the freeze—and stay safe out there!

Dress Smart

The right clothing is your first defense against freezing temperatures. Think of your outfit like a survival shield—each layer plays a role:

- **Base Layer (Stay Dry!)** – This layer touches your skin and should pull sweat away to keep you dry. Skip cotton—it holds moisture and makes you colder. Wet clothes steal heat fast!
  - **Examples:** Thermal long-sleeve shirt, synthetic or wool base layer, moisture-wicking leggings.
- **Insulating Layer (Stay Warm!)** – This is your warmth trapper. Fleece, wool, or puffy materials work best.
  - **Examples:** A fleece hoodie, wool sweater, or insulated puffer vest.

- **Outer Layer (Stay Protected!)** – Shields you from wind and moisture. This should be water-resistant but removable in case you need to cool down.
    - Examples: A waterproof jacket, windbreaker, or heavy winter coat depending on conditions.

Dressing in layers lets you adjust to the temperature, so you don't sweat too much or get too cold. Want to test your setup? Try stepping outside for a few minutes—if you feel too warm or chilly, tweak your layers before heading out!

Keep Moving–But Not Too Much

Moving keeps your body temperature up, but there's a catch—sweat is dangerous in the cold. If you overheat and start sweating, that moisture will make you lose heat even faster.

- **Pace yourself** to stay warm without sweating. If you feel yourself heating up, take off a layer before you get sweaty.

- **Keep breaks short.** The longer you sit still, the faster you'll lose heat. If you need to rest, find a sunny spot and sit on something insulated—never directly on the ground!

Hands, Feet, and Head–Lock in the Heat!

Your body focuses on keeping your core warm, so your hands, feet, and head are the first to get cold. If they're freezing, the rest of you will follow.

- **Gloves:** Always carry two pairs. When one gets too cold, swap it out for a warm pair stored in your pocket.
- **Socks & Boots:** Keep your feet dry! Wet socks = frozen toes. Wool socks work best, and waterproof boots are a must.
- **Hats:** A warm beanie traps heat and keeps your head insulated. If you start sweating, you can easily remove it to cool down before putting it back on when needed.

Stay Hydrated and Fueled

Cold weather tricks you into thinking you don't need water, but dry air pulls moisture from your skin, leading to dehydration. Take small sips of water every few hours to stay hydrated.

- Eat high-energy foods like cheese, nuts, and energy bars to keep your body fueled. Warm meals are best, but calorie-dense snacks will help keep your internal furnace burning.

Fast Warm-Up Tricks

Need to warm up quickly? Try these survival hacks:

- **Rub your wrists together.** Blood vessels near the surface will warm up and help heat your hands.
- **Find a sunny spot.** Standing in the sun—even on a cold day—feels warmer than the shade.
- **Use hand warmers.** Some are disposable, while others can be refueled—just be sure to ask an adult for help with those.

Staying warm isn't just about comfort—it's about survival. With the right clothes, smart movement, and a few key tricks, you can outlast the cold and keep your body from turning into a popsicle!

Beating the Heat – How to Stay Cool When the Sun is Scorching

Blazing sun and high temperatures can drain your energy just as fast as freezing weather. Overheating can leave you exhausted, dehydrated, and even in danger of heatstroke. But don't sweat it! With the right tricks, you can beat the heat and stay cool outdoors.

Dress for the Heat

Your clothing can be your best defense against the sun's relentless rays.

- **Cover Your Head:** A lightweight, breathable hat shields you from the sun and helps keep you cool. Wide-brimmed hats are best, but even a damp bandana under your hat can add extra cooling power.

- **Wear Loose, Light Clothing:** Tightly woven fabrics block the sun's rays, preventing burns, while loose-fitting clothes allow airflow and help sweat evaporate faster. Long sleeves may seem too warm, but if they're lightweight, they can actually keep you cooler while protecting your skin.

Stay Hydrated

Water is your body's built-in cooling system, so drink up!

- **Drink Consistently:** Don't wait until you're thirsty—sip water throughout the day to prevent dehydration.
- **Replenish Electrolytes:** Sweating makes you lose important minerals, so eat salty snacks or drink electrolyte-infused beverages to stay balanced.

Use Water to Cool Down

Water isn't just for drinking—it can cool you down from the outside too.

- **Soak a Bandana:** A wet bandana around your neck or under your hat can work wonders.
- **Dampen Your Hair:** A little water on your head helps with evaporative cooling.
- **Avoid Ice-Cold Water:** Freezing water can shock your system and make your body hold onto heat instead of cooling down. Stick to cool water instead.

Seek Shade and Time Your Breaks

The sun is strongest in the middle of the day, so work smarter, not harder.

- **Take Breaks in the Shade:** Resting in the shade lets your body cool down without working as hard.

- **Avoid Direct Sun Exposure:** Whenever possible, walk in shaded areas or create your own shade with an umbrella or tarp.

Protect Your Skin from the Sun

Sunburn doesn't just hurt—it makes it harder for your body to stay cool. Here's how to shield yourself from harmful rays:

- **Cover Up:** Loose, long-sleeved clothing and wide-brimmed hats provide great protection.

- **Find Shade:** Trees, rocks, or any natural cover can help protect your skin from direct sun exposure.

- **Use Natural Barriers:** No sunscreen? Try mud or certain plant-based coverings (just make sure they're safe for your skin).

- **Time Your Sun Exposure:** If you're in a survival situation, avoid walking during peak sun hours (typically midday) to reduce the risk of overheating and sunburn.

Weird but Effective Cooling Tricks

Beyond the basics, some unexpected tricks can help cool you down even faster!

- **Cool Your Hands, Not Just Your Face:** Soaking your hands in cool water for 15–20 minutes cools your whole body faster than splashing your face.

- **Drink Something Warm:** It might seem backward, but warm drinks can actually trigger your body to sweat more—helping you cool off as the sweat evaporates.

Pro Tip: Feeling dizzy, weak, or confused in the heat? STOP moving and cool down immediately! Find shade, drink water, and use cooling techniques to prevent heat exhaustion from turning into heatstroke.

With these tricks up your sleeve, you'll be ready to take on the heat without breaking a sweat!

Final Thought

Weather can be unpredictable, but the more you pay attention to the world around you, the better you'll get at spotting clues about what's coming. And when you're prepared, no storm or sudden change can catch you off guard—you'll be ready to thrive in any conditions!

Activity 1: Cloud Gazing—What Will the Sky Tell You?

The sky is full of clues—all you have to do is watch the clouds closely and learn what they can tell you about the weather. In this activity, you'll use your super observation skills to study the sky and uncover the secrets hidden in the clouds.

What You'll Need:

- Paper and pencil
- A comfy spot to lie down and gaze up at the sky
- Sunglasses to protect your eyes

Now, get comfortable and start watching!

Clouds are more than just fluffy shapes in the sky—they're like a weather report floating overhead. Some warn you that a storm is coming, while others signal that clear skies are on the way. But before you can start spotting weather clues in the clouds, you need to know what you're looking at!

What to Do:

1. Lie down, relax, and start watching the sky.
2. Identify different clouds and sketch them in your notebook.
3. Predict the weather based on what you see! Did you spot storm clouds? Puffy fair-weather clouds? Keep a record and check later to see if your predictions were right!

Don't worry if you're not sure what type of cloud you're looking at, it takes practice! Just do your best, make a guess, and over time, you'll get better at spotting the clues in the sky. With a little practice, you'll be spotting sky clues and making forecasts with confidence!

How Did You Do?

Now that you've studied the clouds, how well did you predict the weather?

- **Sky Watcher (You guessed 1–2 clouds but weren't sure)** – Great start! Learning to read the sky takes time. Keep watching, and you'll get better at spotting cloud clues!

- **Cloud Reader (You confidently named at least 2 clouds, drew them, and made a weather prediction)** – Awesome! You're starting to see how clouds give hints about the weather. Keep practicing!

- **Meteorologist in Training (You named more than 2 clouds, drew them, guessed the weather—and got it right!)** – Incredible! You've got real weather-watching skills. Keep it up!

Pro Tip: Make a Cloud Viewer! Want to make cloud watching even easier? Cut a rectangle out of cardboard to create a simple sky-viewing frame. Then tape small pictures of different cloud types around the edges, with their names underneath. Hold it up to the sky to help match the clouds you see—it's like having a mini weather lab in your hands!

Activity 2: Making a Barometer

Now you know your clouds—but did you know that changes in air pressure can also help predict the weather? A barometer is a tool that measures air pressure and gives clues about upcoming weather. Meteorologists rely on barometers to make forecasts, but you don't need fancy equipment to see air pressure in action—you can make your own!

What You'll Need:

- A small coffee can or jar
- Plastic wrap
- A rubber band
- A straw, pencil, or lightweight stick
- Tape
- Paper or an index card
- A pencil
- Scissors

Instructions:

1. **Seal the container** – Stretch the plastic wrap tightly over the open end of the jar or can. Secure it in place with a rubber band, making sure it forms a tight seal.

2. **Add the pointer** – Lay the straw, pencil, or stick flat across the plastic wrap so that about two-thirds of its length rests on top, with the remaining third extending past the edge. This setup allows for movement, making it easier to see changes in air pressure.

3. **Tape it down** – Use a small piece of tape to secure the pointer in place on the plastic wrap, but don't press too hard—it needs to move freely with the air pressure changes.

4. **Attach the paper** – Tape a piece of paper or an index card vertically to the back of the container so that part of it extends above and below the pointer.

5. **Mark the starting point** – Look at the pointer from the front and draw a line on the paper where it's currently pointing.

6. **Observe changes** – Check the pointer after 15 minutes and mark any movement. Continue watching over time, adding new marks as needed.

Why Does It Work?

The plastic wrap acts like a flexible surface that responds to changes in air pressure inside the container:

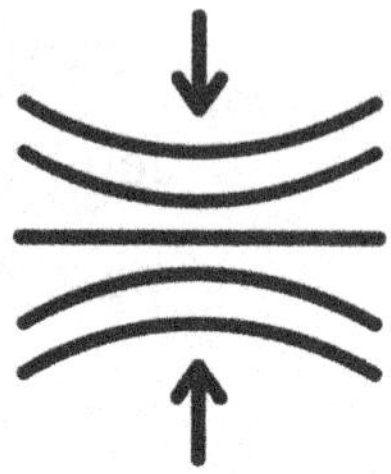

- **When air pressure rises,** it pushes down on the plastic wrap, making the pointer tilt up.
- **When air pressure drops,** the plastic wrap expands slightly, making the pointer tilt down.

Since falling pressure often means stormy weather and rising pressure signals clear skies, your homemade barometer can help you predict what's coming!

A homemade barometer gives you a rough idea of air pressure changes, but for a more detailed forecast, check a weather report.

Activity 3: Daily Weather Watcher—Can You Predict What's Next?

We know weather forecasts aren't always right—but many people rely on them to plan their day. The good news? You don't need a fancy forecast to get prepared—you just need your own eyes, ears, and a little practice! In this activity, you'll use what you've learned to make your own weather predictions. As you keep watching and practicing, you'll be able to notice patterns and prepare for whatever the day brings!

What You'll Need:

- A notebook or paper
- A pencil
- A thermometer (if available)
- A window with a good view—or better yet, step outside!

Step 1: Observe and Record

Each day, take a few minutes to carefully observe what's happening outside. Use all your senses—what do you see, feel, hear, and smell? Here's what to write down:

1. **Cloud Type & Color:** Are the clouds fluffy, wispy, dark, or streaky?
2. **Precipitation:** Is it dry, drizzling, raining hard, snowing or hailing?

3. **Wind:** Is it calm, breezy, or strong? Which direction is it blowing?

4. **Temperature:** Does it feel warm, cool, or freezing? (Use a thermometer if you have one!)

5. **Other Clues:** Can you smell rain in the air? Hear distant thunder?

Step 2: Make Your Prediction!

Based on what you saw, what do you think will happen next?

Will it rain? Get windier? Will the sky clear up?

Step 3: Check Your Guess

Later in the day, compare your prediction to the actual weather—or check a weather app. How close were you? Keep practicing to sharpen your weather-watching skills!

Try It for 3–5 Days!

Do this activity each day for at least 3 to 5 days in a row. You'll start noticing patterns and get better at predicting what's coming. How accurate can your guesses get by the end of the week?

Bonus Activity: Make a Simple Rain Gauge

Want to take your weather tracking to the next level? Try making a rain gauge!

You'll need:

- A clear plastic bottle or jar
- Scissors
- A ruler
- A marker
- A few small stones or pebbles

How to Make It:

1. If using a plastic bottle, cut the top off the bottle about 3–4 inches down, and then flip the top upside down and place it inside the bottom like a funnel. If using a jar, no cutting needed!

2. Add a few small stones to the bottom to keep it from tipping over.

3. Use the ruler and marker to draw measurement lines starting at the bottom.

4. Place your rain gauge outside in an open area. After it rains, check how much water collected!

Why It's Cool: You'll see how much rain actually falls instead of just guessing. Compare it to weather reports to see how accurate your measurements are!

Wilderness Warrior Skill Unlocked!

You've mastered how to predict and survive wild weather! Mark your progress by coloring in your Wilderness Warrior Weather Master Award on your Survivalist Skill Tracker (page 278).

Level Up Your Survival Skills!

You've learned how to predict the weather and stay safe—rain or shine. But what if there's no food in sight? It's time to learn how to forage for food—without getting sick!

Up next: Skill 6 – Foraging! Let's GO!

Survival Skill 6: *Foraging*

"Wild foods are not only free, they are superior in flavor and nutrition." — Euell Gibbons

Foraging Safety Rules

You've got shelter. You've got fire. You've got water. But now your stomach is growling.

Here's the deal—you can survive up to three weeks without food, but that doesn't mean you'll enjoy it. Hunger makes you weak, tired, and cranky (and no one likes a hangry survivalist!).

The good news? Nature is like a giant grocery store... IF you know what to look for. The bad news? Pick the wrong thing, and you could get sick (or worse). That's why foraging isn't just about finding food—it's about knowing what's safe to eat and what to avoid.

Foraging Like a Wilderness Warrior

Foraging is like a treasure hunt where the prize is food, but before you start snacking on wild berries and mystery leaves, follow these golden rules:

1. **Know Where You Can Forage** – Not all land is fair game for gathering food. Some places require a permit, and others don't allow foraging at all. Always check the rules with an adult before you start picking.

2. **Avoid Polluted Areas** – Would you eat food off the ground at a gas station? Nope! The same goes for plants growing near roads, factories, or places where chemicals are used. Stick to clean, natural areas.

3. **Take Only What You Need** – Many wild foods spoil fast, and animals depend on them too. Take just enough for a meal and leave the rest for nature's other hungry guests.

4. **Never Forage Alone** – Foraging is way more fun with a buddy—and way safer. Always go with an adult or someone who knows plants well. If you get lost or eat

something you shouldn't, you'll want help nearby!

5. **Choose Healthy Plants** – If a plant looks sick, moldy, or just plain weird, skip it! You wouldn't buy rotten fruit at the store, so don't pick it in the wild.

6. **Test for Allergies** – Trying a new wild food? Start small. Place it on your lips first—if you feel itching, burning, or swelling, spit it out and don't eat it!

7. **NEVER Eat Wild Mushrooms** – Seriously—just don't. Many wild mushrooms are poisonous, and even experts can have trouble telling them apart. If you're curious, take a foraging class with an expert instead of guessing!

Pro Tip: Keep a nature notebook! If you spot an interesting plant, write down what it looks like and where you found it. Later, use a field guide (or a knowledgeable adult) to learn if it's safe to eat. If you're not 100% sure—DON'T TOUCH IT!

You're well on your way to becoming a true Wilderness Warrior! Master these rules, and soon you'll be finding wild snacks with confidence—without any risky mistakes.

Common Edible Plants by Season

Knowing what's available in each season can help you forage successfully. Below is a guide to wild edibles you may find throughout North America. Remember, always be 100% sure of your identification before eating anything!

Foraging Rule #1: **When in Doubt, Stay Out!**

If you're ever unsure about a plant, don't eat it! Some toxic plants look very similar to edible ones. Mistaken identity can lead to severe illness—or worse. Always cross-check with a reliable guide, and when possible, consult an experienced forager.

Happy (and safe) foraging!

Spring Foraging (Fresh greens and early blooms)

Spring is one of the best times to forage for tender greens, young shoots, and early blossoms. Many plants taste best when they are young before they become tough or bitter.

Dandelion (Taraxacum officinale)

Often considered a lawn weed, dandelions are packed with vitamins and minerals.

How to Identify:

- Bright yellow flowers with numerous petals
- Deeply lobed, serrated leaves growing in a rosette at the base
- Hollow stems that leak a white, milky substance when broken (can irritate sensitive skin)

How to Eat:

- Entire plant is edible! Young leaves (early spring) are less bitter. Flowers can be used for teas or even cookies.

Dandelion

Stinging Nettle (Urtica dioica)

Yes, it stings—but once cooked, it's delicious and nutrient-packed, similar to spinach.

How to Identify:

- Tear-drop-shaped, serrated leaves in opposite pairs
- Hairy stems with tiny, stinging spines under the leaves
- Often found in damp areas, forests, and gardens

Caution: Always wear gloves when harvesting—this plant will sting!

How to Eat:

- Boil or steam the leaves to remove the sting.
- Best harvested in spring when under a foot tall for the most tender leaves.

Ramps, Wild Leeks (Allium tricoccum) - AKA Wild Garlic

These tasty plants pop up in the spring, hiding in shady forests with damp soil. If you find them, lucky you—chefs go crazy for these wild greens!

How to Identify:

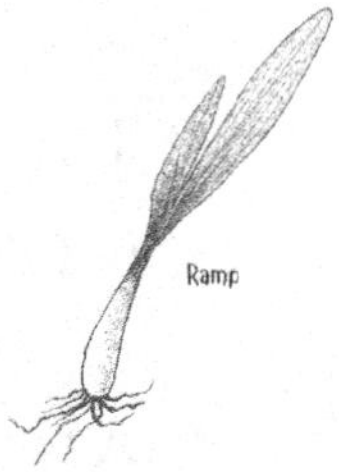

- Smooth, bright green leaves that grow in bunches
- White flowers in clusters (but they don't always bloom)
- Smells like garlic when you crush a leaf—if it doesn't, uh-oh! It could be Lily of the Valley, which is poisonous.

How to Eat:

- Leaves, flowers, and bulbs are all edible, but harvest with care! Ramps grow slowly—only take one leaf per plant so they can regrow.
- Enjoy raw or cooked, but try a small bite first if you're new to them.

Chickweed (Stellaria media)

A mild, tender green that grows almost year-round and is packed with vitamins. Perfect for fresh eating!

How to Identify:

- Small, oval-shaped leaves with a slight point at the tip
- Tiny white star-shaped flowers with five deeply lobed petals
- A single line of tiny hairs running down the stem (a key identifier!)
- Often found in gardens, fields, and shady, moist areas

How to Eat:

- Enjoy raw in salads or sandwiches—it has a mild, slightly sweet flavor.
- Add to soups or sauté lightly like spinach.
- Best harvested when young and tender.

Basswood (Tilia americana)

Also known as linden, this tree is common in forests and even cities. It has mild, edible young leaves that can be a great survival snack!

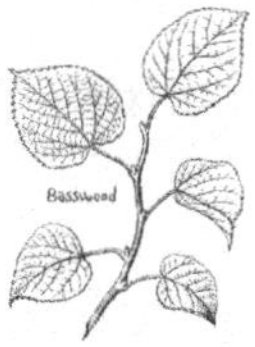

How to Identify:

- Heart-shaped leaves with finely toothed edges
- Smooth, grayish bark on young trees, becoming rougher with age
- Small, fragrant, yellowish-white flowers in clusters (later in summer)

How to Eat:

- Young leaves can be eaten raw in salads or sandwiches.
- The inner bark is technically edible and can be dried and ground into a flour-like powder, but it's rarely used due to its fibrous texture and mild taste.

Violets (Viola spp.)

These delicate wildflowers add a pop of color to woodlands and lawns—and they're completely edible!

How to Identify:

- Heart-shaped leaves with slightly scalloped edges
- Five-petaled flowers, often purple, but can also be white, blue, or yellow
- Low-growing and often found in shady, moist areas

How to Eat:

- Leaves and flowers are both edible and great in salads or as a garnish.
- Flowers can be candied or used to make violet syrup.
- Leaves are mild and can be cooked like spinach.

Clover (Trifolium sp.)

Clover is a common plant found in meadows, fields, and lawns, and it's packed with nutrients. It makes a great addition to teas, salads, and baked goods.

How to Identify:

- Leaves usually have three leaflets, sometimes with a faint white "V" marking.
- Small, round flower clusters that can be white, pink, or red.
    - Red clover (Trifolium pratense) is taller, with reddish-pink flowers.
    - White clover (Trifolium repens) is low-growing and often found in lawns.
- Spreading growth habit, often forming patches.

How to Eat:

- Leaves and flowers can be eaten raw, dried for tea, used in baking, or made into syrup!

Lookalikes:

- Wood sorrel has three-part leaves but is easy to tell apart by its heart-shaped leaves and tangy, lemony flavor.

Wood Sorrel (Oxalis sp.)

A small, delicate plant with heart-shaped leaves, wood sorrel has a bright, lemony flavor that adds a zesty kick to foraged meals.

How to Identify:

- Clover-like leaves, each divided into three heart-shaped leaflets
- Small yellow, white, or pink flowers with five petals
- Often found in shady areas, lawns, and woodland edges

How to Eat:

- Eat fresh as a tart snack, sprinkle on salads, or brew into a refreshing tea.
- Enjoy in small amounts—like spinach, it contains oxalic acid, which can irritate in large quantities, especially for those with kidney issues.

Cattail Shoots (Typha sp.)

Cattails grow in wetlands, marshes, and along the edges of ponds and lakes. The young shoots are tender and edible.

How to Identify:

- Tall, grass-like plant with long, flat green leaves
- Brown, sausage-shaped flower head in late summer
- Found in or near shallow water

How to Eat:

- Peel young shoots and eat raw or cooked (taste similar to cucumber).
- Rhizomes (underground stems) are starchy and can be roasted or ground into flour.

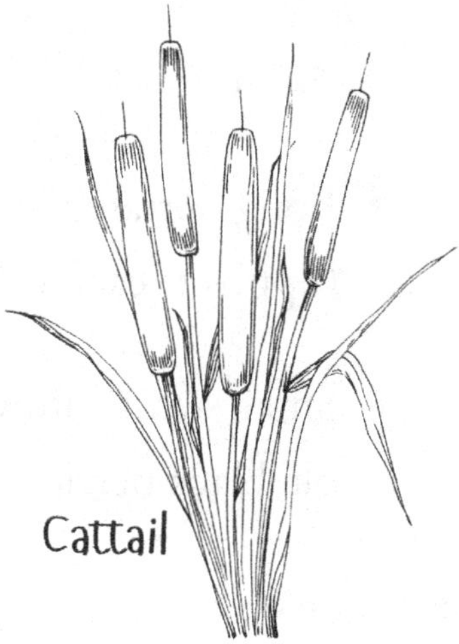

Summer Foraging (Berries, flowers, and more greens)

Summer brings a bounty of berries, flowers, and edible weeds. This is a great time for wild fruit and nutrient-dense greens.

Elderberry (Sambucus sp.)

Elderberry shrubs thrive in light shade and are often found near water sources like ponds and lakes.

How to Identify:

- Bumpy bark
- Long, serrated leaves growing in pairs (opposite each other)
- Small white flower clusters turning into dark purple-black berries on pinkish-red stems

How to Eat:

- **Never eat raw elderberries**—they can cause nausea. Instead, cook or dry them for tea.

Blackberry & Red Raspberry (Rubus sp.)

These berries grow on thorny shrubs and are delicious and nutrient-rich.

How to Identify:

- Thorny brambles with serrated, oval leaves
- Clusters of white, five-petaled flowers
- Berries grow in clusters: Blackberries ripen from red to black, while raspberries stay red

How to Eat:

Eat fresh, dry for later, or cook into jams.

Lookalikes:

- Thimbleberry, Salmonberry, and Dewberry – Edible! These similar berries are safe to eat and taste great.
- Virginia Creeper (Parthenocissus quinquefolia) – Not safe to eat! Its dark berries may look like blackberries but grow in small clusters on a vine with five-parted leaves instead of thorny brambles

Blueberry (Vaccinium sp.)

Wild blueberries love to grow in pine barrens—sandy, nutrient-poor soils where few other plants thrive. But don't let that fool you! These tiny berries pack a punch when it comes to flavor and nutrition.

Blueberries

How to Identify:

- Small bushes with oval, smooth-edged, dark green leaves (turning red in fall)
- White, bell-shaped flowers
- Individual blue to purple berries

How to Eat:

- Eat fresh, dry for later, or cook into recipes.

Fun Fact: Wild blueberries have **twice the antioxidants** of store-bought ones, making them a super-powered snack straight from the forest!

Cattail Pollen (Typha sp.)

By summer, cattails produce bright yellow pollen, which can be harvested and used as a flour substitute.

How to Identify:

- Look for the distinctive brown flower head, which will release bright yellow pollen when shaken.

How to Eat:

- Collect pollen by shaking the flower heads into a bag.
- Use as a flour substitute in baking or mix into soups for extra nutrition.

Purslane (Portulaca oleracea)

A hardy, low-growing succulent with a slightly sour, crunchy taste, purslane is packed with omega-3s and can be eaten raw or cooked.

Purslane

How to Identify:

- Thick, reddish stems with smooth, paddle-shaped green leaves
- Leaves grow in clusters at stem joints and ends
- Small yellow flowers that open in sunlight

How to Eat:

- Enjoy raw in salads for a refreshing crunch, sauté as a leafy green, or add to soups and stews for extra nutrition.

Lamb's Quarters (Chenopodium album)

Sometimes called wild spinach, this nutrient-rich plant is loaded with vitamins and minerals, making it a great foraging find.

Lamb's Quarters

How to Identify:

- Diamond-shaped, toothed leaves with a powdery, whitish coating
- Tall, branching stems that may have a slight reddish hue
- Small greenish flowers that grow in clusters at the top

How to Eat:

- Use fresh in salads, steam like spinach, or add to soups and stir-fries. Best harvested when young and tender.

Fall Foraging (Nuts, seeds, and hardy fruits)

Fall is when trees and plants prepare for winter, making it a great time to collect nuts, seeds, and hardy fruits.

Wild Apple (Malus sp.)

Wild apple trees can sometimes be found in forests or old homesteads.

How to Identify:

- Short trees with oval, serrated leaves
- Thin, red-brown to gray-brown bark with shallow grooves
- White flowers with pink centers and five petals

How to Eat:

Fruit is edible but may be sour or bitter. Cooking enhances flavor.

Cranberry (Vaccinium macrocarpon)

Cranberries love wet, boggy places and are one of the few fruits that ripen late in the year—sometimes even surviving under a blanket of snow! These bright red berries are super tart, but if you find them after the first frost, they get sweeter.

How to Identify:

- Low-growing plants with small, glossy, oval leaves
- White to pink flowers with backward-turned petals
- Small, round red berries (often clinging to the plant even after frost)
- Fresh cranberries are super sour but packed with vitamin C—try them raw if you like a zingy snack!
- Dry them for trail mix, steep them in tea, or cook them into jams and sauces.

Rosehip (Rosa sp.)

After a wild rose loses its petals, it leaves behind a secret superfruit—rosehips! These small, bright fruits are bursting with vitamin C and were even used as a survival food during wartime.

How to Identify:

- Found on wild rosebushes after flowers drop.
- Oval, smooth, or slightly wrinkled fruits that turn red to deep orange when ripe.
- Bushes have thorny stems and serrated green leaves.

How to Eat:

- Best after the first frost! Cold temperatures make them softer and sweeter.
- Cut open and remove seeds and fuzzy hairs (they can irritate your mouth).
- Eat fresh, dry for tea, or make into jams and syrups for a tangy boost of nutrients.

Acorns (Quercus sp.)

A staple survival food, acorns are packed with nutrients but must be processed to remove bitter tannins before eating.

How to Identify:

- Hard, smooth nuts enclosed in a rough-textured cap
- Grown on oak trees with lobed or toothed leaves
- Found on the ground in autumn after falling from the tree

How to Eat:

- Shell the acorns and soak or boil them in multiple changes of water to remove the bitter taste. Keep going until the tannins (which make the water turn brown) are gone and the water runs clear. Once processed, you can grind them into flour for baking or roast them for a crunchy, nutty snack!

Hickory Nuts (*Carya* sp.)

Closely related to pecans, hickory nuts are a high-energy wild food packed with healthy fats.

How to Identify:

- Round or oval nuts with a thick, hard shell that splits open when ripe
- Found on hickory trees with compound leaves made up of 5–9 leaflets
- Bark can be smooth or slightly ridged, depending on the species

How to Eat:

- Gather nuts after they fall to the ground in autumn.

- Crack open the shell and remove the nut inside.
- Some species, like **shagbark** and **shellbark**, have sweet, edible nuts. While others, like **bitter hickory**, live up to their name—not tasty at all!

Walnuts (Juglans sp.)

All kinds of wild walnuts are edible, and they're packed with nutrients—but you'll need a little patience to harvest them!

How to Identify:

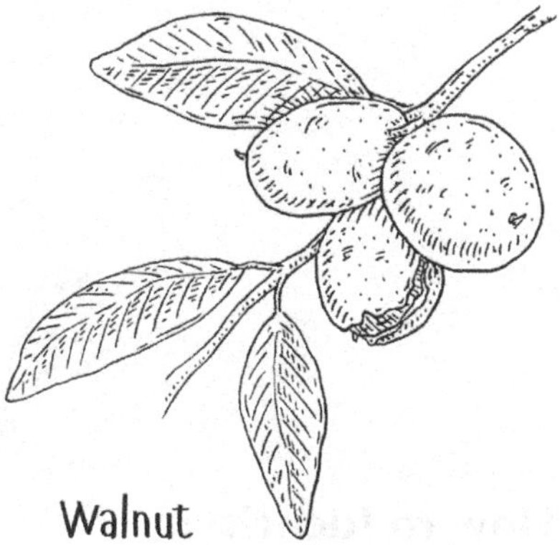

Walnut

- Large green husks encasing the hard shell
- Grown on walnut trees with pinnate leaves (multiple leaflets on a central stem)
- Leaves have a strong, earthy scent when crushed

How to Eat:

- Pick walnuts when they fall to the ground and allow the green husks to darken and soften for one to two weeks in a dry, well-ventilated area. Once softened, remove the husks, wash and dry the nuts, then crack them open to eat raw or toast for extra flavor.

Basswood Seeds (Tilia sp.)

Sometimes called "wild popcorn," these small seeds from the basswood tree can be a fun foraged snack.

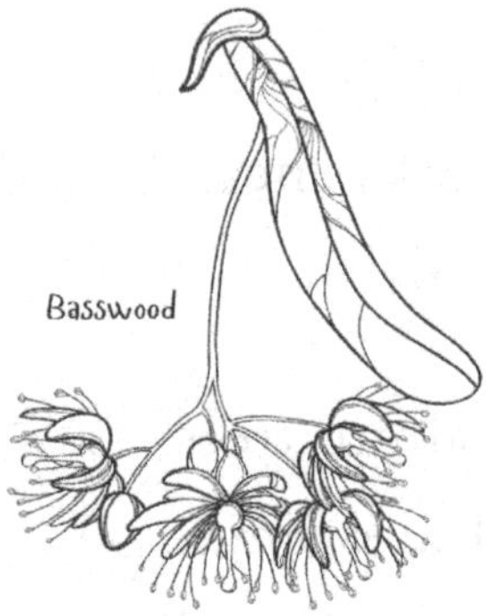

How to Identify:

- Small, round seeds attached to a papery bract that helps them drift in the wind
- Found on basswood trees with heart-shaped leaves and smooth gray bark
- Creamy-yellow flowers that are fragrant and attract bees

How to Eat:

Crack open the seeds and eat them raw or roast them for a crunchier texture. Young leaves of basswood trees are also edible and can be eaten in salads.

Wild Grapes (Vitis sp.)

These trailing vines produce small, tart grapes that can be eaten fresh or made into juice or jelly.

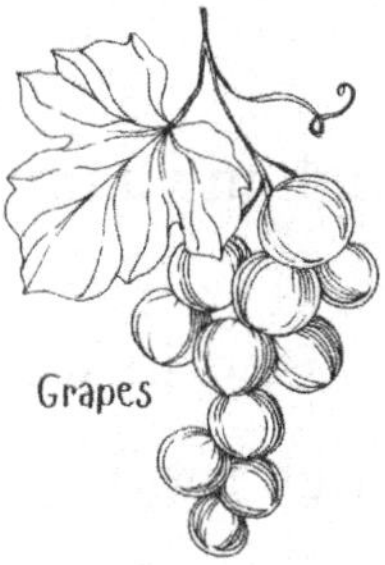

How to Identify:

- Woody vines that climb trees and shrubs, with curly tendrils
- Large, lobed leaves with serrated edges
- Small clusters of dark purple or blue grapes in late summer to fall

How to Eat:

- Eat fresh, though they are often tart. Use in jams, jellies, or dry them into raisins. Be sure to avoid lookalikes like Canada Moonseed, which is toxic (moonseed has smooth-edged leaves and a single large seed inside).

Winter Foraging (Cold-hardy plants and survival foods)

Foraging in winter is tough, but some foods can still be found if you know where to look.

- **Pine Needles** – High in vitamin C; steep for tea.
- **Cattail Roots** – Found in wetlands; can be boiled or roasted for energy-rich food.
- **Rosehips** – Hardy enough to last through the first snow; sweeter after frost.
- **Inner Bark (Basswood, Pine, Birch)** – The soft inner bark can be dried and ground into flour.
- **Frozen Cranberries** – Sometimes found clinging to plants under the snow.

Pine Branch

Top 10 Poisonous Plants to Avoid

Foraging is an exciting way to connect with nature and find wild foods, but not every plant is safe to touch or eat. Some plants can cause rashes, while others can make you very sick—or even be deadly. You might have heard of poison ivy, but many other plants are just as dangerous. Some even look like safe plants, making them tricky to tell apart. **If you aren't 100% sure what a plant is, the best rule is: DO NOT TOUCH or EAT it.**

Here are 10 poisonous plants you should always avoid:

1. Queen Anne's Lace

How to Spot It:

- Tightly packed white flowers (not loose like hemlocks)
- Feathery, deeply lobed leaves (like carrots)
- Fuzzy stem with light grooves

* **Danger:** Not deadly, but can cause skin irritation if you have sensitive skin.

2. Giant Hogweed

How to Spot It:

- Grows 6–18 feet tall
- Huge umbrella-shaped clusters of tiny white flowers
- Thick, hollow stem with ridges or purple spots
- Giant, deeply lobed leaves that look like oversized pumpkin leaves

* **Danger:** The sap can cause severe blisters and burns when exposed to sunlight. If it gets in your eyes, it can cause temporary or permanent blindness!

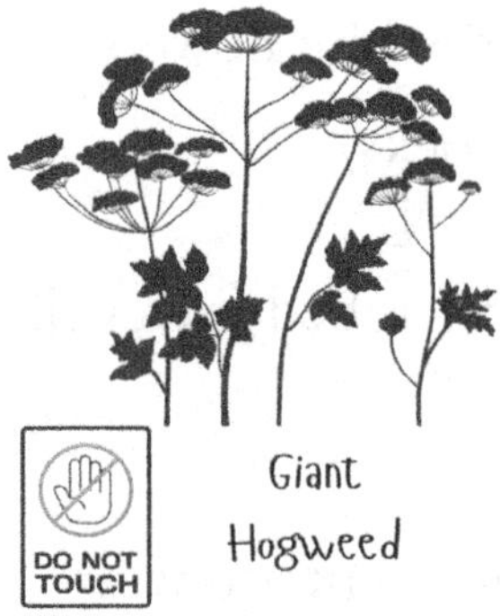

3. Poison Hemlock

How to Spot It:

- Umbrella-like white flower clusters (similar to Queen Anne's lace)
- Feathery green leaves that look like carrot tops
- Grows 3–8 feet tall
- Hairless, hollow stem with purple blotches

* **Danger:** Every part of this plant is deadly if eaten—even a tiny amount can kill humans and animals. Even touching these plants can transfer toxins to your skin. Always wash your hands if you come into contact with them!

4. Water Hemlock

How to Spot It:

- Grows 3–6 feet tall, often near wetlands or streams.
- Small white flowers in umbrella-shaped clusters.
- Leaves are more elongated than poison hemlock and sharply serrated.
- Stems are NOT blotchy but have hollow chambers inside.

* **Danger:** Considered one of the most poisonous plants in North America—eating it will kill you!

Pro Tip: If it looks like a wild carrot or Queen Anne's lace but has purple blotches on the stem (Poison Hemlock) or a hollow, chambered stem (Water Hemlock), STAY AWAY!

5. Cow Parsnip

How to Spot It:

- Large maple-shaped leaves with deep lobes
- Hairy, grooved stems
- Grows 4–10 feet tall
- White flower clusters form an umbrella shape

* **Danger:** Touching it can cause painful rashes and blisters, similar to giant hogweed.

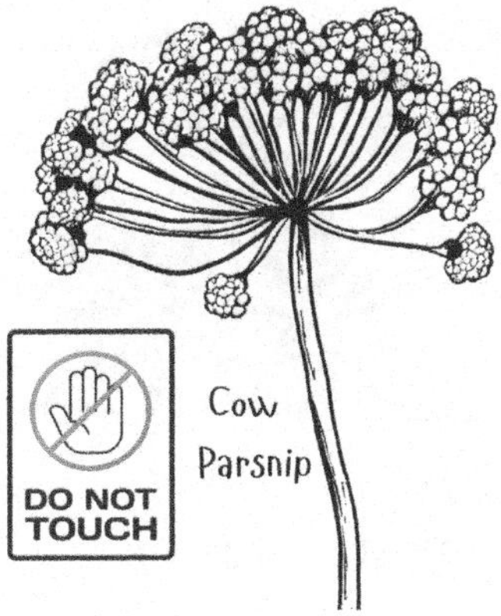

6. Wild Parsnip

How to Spot It:

- Looks like cow parsnip, but less hairy
- **Yellowish-green** flower clusters form loose umbrellas
- Lightly hairy, grooved stems
- Grows 2–5 feet tall

* **Danger:** The sap reacts with sunlight, causing severe blisters, burns, and skin discoloration.

Pro Tip: Feeling overwhelmed by all these plants? Don't worry! Even expert foragers had to start somewhere. The key is to slow down and pay close attention to the details—especially the stems, leaves, and flower clusters. Here's how to tell them apart:

Queen Anne's Lace – Small white flowers packed closely together. Fuzzy green stems. Safe but can irritate skin.

Poison Hemlock – White flowers in loose umbrella clusters, smooth stems with purple blotches. Deadly if eaten!

Water Hemlock – Similar flowers, but hollow, chambered stems and grows near water. Most toxic plant in North America!

Cow Parsnip – Large, deeply lobed leaves, hairy stems, and umbrella-shaped flowers. Causes rashes.

Giant Hogweed – HUGE plant, thick ridged or purple-spotted stems, massive leaves. Causes severe burns and blindness!

7. Morning Glory

How to Spot It:

- Bright, trumpet-shaped flowers that open in the morning
- Heart-shaped leaves
- Vine-like growth that creeps along the ground

* **Danger:** The seeds are toxic and can cause severe stomach pain, numbness, and muscle spasms if eaten.

8. Poison Oak

How to Spot It:

- Leaves of three, let it be!
- Leaves are fuzzy, oval-shaped, and serrated
- Green in summer, red in fall
- Produces yellow-white berries

* **Danger:** Touching it causes itchy rashes—never bring it near your face!

9. American Pokeweed

How to Spot It:

- Bright pink stems with purple-black berries
- Long, oval, smooth-edged leaves
- Clusters of small white-green flowers before berries appear

* **Danger:** The berries may look tempting, but they are poisonous to humans! Eating them can cause vomiting, breathing problems, and convulsions (uncontrollable shaking).

10. Yew

How to Spot It:

- Evergreen tree or shrub with soft, flat, needle-like leaves
- Bark is dark brown to reddish and scaly
- Produces small, cup-shaped red berries with a single seed inside

* **Danger:** All parts of the yew tree—especially the seeds inside the berries—are highly toxic. Eating them can cause dizziness, irregular heartbeat, severe illness, or even death. No matter what, **yew should never eat yew**!

Wild Foraging Adventure: Learn, Research, and Taste!

Before you start gathering wild foods, you need to be 100% sure what you're picking. Some delicious plants have toxic twins—so let's train your foraging skills first!

Activity 1: The Mushroom Mystery Challenge – JUST SAY NO!

Wild mushrooms are one of the riskiest foods to forage. Some look nearly identical to deadly ones. Even expert foragers use guidebooks and double-check their finds with professionals—mistakes can be fatal!

THE RULE? We will NOT be eating wild mushrooms in this challenge. Learning about them is great, but safely foraging them takes years of practice with expert guidance.

Your Challenge:

- Research at least three edible wild mushrooms in your area and their poisonous lookalikes.
- Write down how they are the same and how they are different—pay close attention to color, shape, gills, and stem features!
- **Go the extra mile:** Draw a Mushroom Match-Up diagram to compare them.

Remember: Even the pros can make mistakes. That's why we won't be eating wild mushrooms in this activity. The morel of the story? If

you're ever interested in learning to forage mushrooms safely, find an experienced mentor to guide you!

How Did You Do?

- **Mushroom Newbie** – You found 1 mushroom and its lookalikes. Great start! Keep training your eyes!
- **Fun-Guy Researcher** – You found 3 mushrooms and their toxic twins. You're getting pretty sharp!
- **Master Mycologist** – You found 3+ mushrooms, drew them, and labeled their key differences. Look at you, mushroom detective!

Activity 2: The Great Plant Lookalike Investigation

Now that you've trained your eyes on mushrooms (but remember, we're NOT eating them!), let's test your skills on wild edible plants! Just like mushrooms, many tasty plants have toxic copycats—so knowing the difference is a MUST.

Your Challenge:

- Look up three edible wild plants in your area and their poisonous lookalikes.
- Write down how you can tell them apart. What key features do you need to watch for?
- **Extra credit:** Draw them! Label their unique characteristics to create your own foraging guide.

How Did You Do?

- **Leaf Peeker** – You found 1 edible plant and its lookalike. Not bad for a first forage!
- **Plant Sleuth** – You found 3 edible plants and their

toxic twins. You're becoming a foraging pro!

- **Wild Edible Expert** – You found 3+ plants, drew them, and labeled their features. You're ready to teach a class!

Activity 3: The Wild Kitchen Challenge – Safety First!

Now that you've done your research, it's time to put your knowledge to the test! If you can safely forage (with an adult's permission), try gathering ingredients for one of these recipes:

1. **Wild Garlic Butter** – Easy and delicious! If you don't have wild garlic, store-bought works too.

2. **Dandelion & Lemon Cookies** – A fun way to turn a "weed" into a tasty treat!

3. **Stinging Nettle Soup** – Handle with care! If you can't find nettles, try using spinach instead.

Foraging Safety Reminders:

- Only pick plants if you are 100% sure they are safe.

- Never eat plants that could be contaminated by pesticides, fertilizers, or vehicle exhaust.

- Wash all foraged foods before eating.

- When in doubt, leave it out!

Safety First!

Always have an adult's permission and assistance when using kitchen tools like knives, processors, pans, and the oven. Be extra careful when foraging—make sure you correctly identify wild plants before using them in recipes. If you're unsure, ask an expert!

No safe foraging spots near you? No problem! Each recipe has an alternative ingredient so you can still join the challenge.

Wilderness Warrior Skill Unlocked!

You've mastered how to safely find, identify, and prepare your foraged food! Mark your progress by coloring in your Wilderness Warrior Foraging Master Award on your Survivalist Skill Tracker (page 278).

Level Up Your Survival Skills!

Now you know how to find food in the wild! But what if you wander too far and can't find your way back? It's time to learn how to navigate like a true explorer!

Up next: Skill 7 – Navigating the Wilderness! Let's GO!

Wild Garlic (Ramps) Butter

Ramp up your garlic butter with this wild and delicious spread—perfect for bread, buns, seafood, veggies, you name it!

Time: 10 minutes

Makes: About 9 oz

Ingredients:

- 3 ½ oz (about 1–1½ cups) wild garlic (or use spinach + 1 minced garlic clove or 1 tbsp garlic powder for a milder taste)
- 1–2 tbsp fresh lemon juice
- 9 oz butter (softened)
- ¼ tsp black pepper
- 1–1 ½ tsp salt (to taste)

Instructions:

Food Processor Method (Easiest!):

1. Add all the ingredients into a food processor.
2. Blend until smooth.

No Food Processor? No Problem!

1. Finely chop the wild garlic (or spinach).
2. In a bowl, mash the butter with a fork.
3. Add the chopped greens, lemon juice, salt, and pepper.
4. Keep mashing and stirring until everything is well mixed.

Tips:

- **Butter should be at room temperature**—this makes it easier to mix.
- **Season carefully!** Start with a little salt and pepper, then taste and add more if needed.
- **Too strong?** Use half the wild garlic and see what you like.
- **Storage:** Your butter will last up to a week in the fridge. For longer storage, wrap it in plastic wrap and shape it into a log or store it in an airtight container in the freezer for up to 3 months.

Dandelion & Lemon Cookies

Never think of dandelions as weeds again after trying these bright, citrusy cookies!

Time: 25 minutes

Makes: About 12 cookies

Ingredients:

- 1 ½ cups all-purpose or whole wheat flour (6.4 oz)
- 20 dandelion flower heads (just the yellow petals)
- ½ cup sugar (3.5 oz)
- Zest of 1 medium lemon
- ½ cup butter, softened (4 oz)
- 1 large egg
- 1 tbsp fresh lemon juice
- Pinch of salt

Instructions:

Step 1: Prepare the Dandelions

- Pick fresh dandelions and soak them in water overnight to remove bugs.
- Drain, pat dry, and pluck the yellow petals from the flowers.

Step 2: Make the Dough

- Preheat the oven to 350°F and grease a baking sheet.
- In a bowl, mix the butter and sugar until creamy.
- In another bowl, whisk the egg and lemon juice, then mix it into the butter mixture.
- Stir in the lemon zest.
- Slowly add the flour and a pinch of salt, mixing well.
- Fold in the dandelion petals.
- Step 3: Bake the Cookies
- Scoop out tablespoon-sized portions and place them on the baking sheet, leaving space between each.
- Bake for 13–16 minutes, until lightly golden.
- Carefully remove and cool on a wire rack.

Tips:

- **Store cookies in an airtight container for up to a week.**
- **No dandelions? Try using other edible flowers like lavender, calendula, or rose petals!**

Stinging Nettle Soup

Turn this wild plant into a creamy, nutritious soup!

⚠ Always wear gloves when handling raw nettles!

They sting!!!

🕒 Time: 42 minutes

🍴 Serves: 6

Ingredients:

- 8 oz fresh nettle tops (about ½ a large shopping bag)
- 1 lb potatoes, peeled & chopped
- 1 tsp butter
- 1 tbsp olive oil
- ½ cup celery, chopped
- ½ cup onions or shallots, chopped
- 1 tsp dried thyme
- 4 cups chicken or vegetable broth
- 1–2 cups water
- 1 bay leaf
- 1 ½ tbsp lemon juice
- 2–3 tbsp heavy cream (optional)
- Salt & pepper to taste

Instructions:

Prepare the Nettles:

1. Bring a pot of salted water to a boil.
2. Using gloves, add nettles and blanch (boil) for 2 minutes.
3. Transfer to ice water, then drain.
4. Trim leaves from stems—you'll need about 3–4 cups of leaves.

Make the Soup:

1. Heat butter & olive oil in a pot over medium heat.
2. Sauté onions or shallots & celery for 5 minutes until soft.
3. Add potatoes, thyme, bay leaf, and broth. Simmer 5 minutes.
4. Add chopped nettles and enough water to cover everything. Simmer 15 minutes.
5. Remove bay leaf, then blend soup until smooth.
6. Stir in lemon juice, salt, and pepper to taste.
7. Serve with a swirl of cream if desired.

Survival Skill 7:

Navigation

"Not all those who wander are lost... but if you don't know where you're going, you might be." — J.R.R. Tolkien

Mastering the Compass: Let It Point You Home!

Imagine this: You're deep in the wilderness. The sun is setting, your phone has no signal, and every tree looks the same. Do you panic? Nope—because you're about to become a navigation master!

Getting lost in the wild is no joke. If you don't know how to find your way, you could spend hours—or even days—wandering in circles. But here's the good news: you don't need a GPS or a high-tech map to navigate! For thousands of years, explorers, adventurers, and survivalists have found their way using the sun, stars, and nature itself. Ready to give it a try?

With the right skills, you'll always know where you are, where you started, and how to get where you need to go. A map and compass can be lifesavers—but only if you know how to use them. And don't count on your phone's GPS to bail you out. Batteries die, signals drop, and screens crack. That's why every Wilderness Warrior should master navigation—with AND without technology

In this chapter, you'll unlock the secrets of reading maps, using a compass, and tracking your path like a real explorer. Pay attention, practice often, and soon you'll have the confidence to find your way anywhere—even if you're dropped in the middle of nowhere.

One of the most important tools in your navigation toolkit is a map. But not just any map—you need the right one for the job!

How to Use a Map

Not all maps are created equal! Some help you find the best pizza place in town, while others can guide you safely through the wild. But if you're lost in the middle of nowhere, you need the right kind of map—or you might just end up going in circles!

Types of Maps: Which One Do You Need?

- **Topographic Map** – Your ultimate wilderness survival tool! This map is packed with details like mountains, rivers, forests, and even elevation (how high things are above sea level). It's designed to be used with a compass, so you can find your way even without GPS.

- **Road Map** – Perfect for road trips, not survival. It shows streets, highways, and major buildings like hospitals. Handy in a car—not so much in the deep woods!

- **Tourist Map** – If you're looking for cool attractions, this is the one. It highlights hotels, landmarks, and points of interest—but don't count on it to help if you're lost in the wild!

Survival Rule: If you're heading into the wilderness, always bring a topographic map of the area. It's the only one that can help you get back to safety!

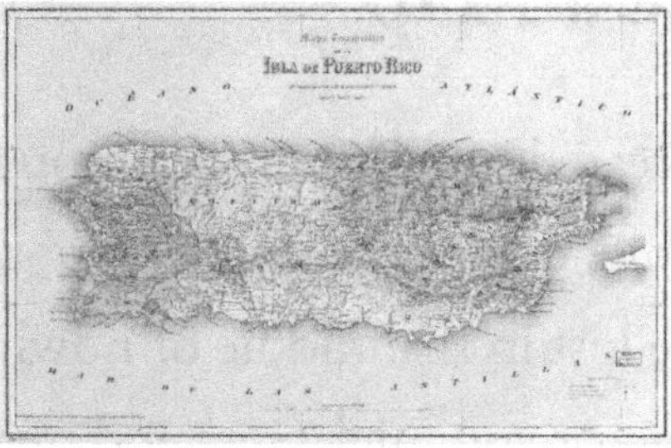

Cracking the Map Code

At first glance, a topographic map might look like a jumbled mess of squiggly lines and weird symbols. But don't worry! Every piece of information is a secret code that helps you find your way. Let's break it down:

- **Map Scale** – You'll see numbers like 1:50,000. This means that one inch on the map equals 50,000 inches (or almost 0.8 miles) in real life. Different maps have different scales, so always check!

- **Legend (Key)** – This is like the map's secret decoder ring. It explains all the symbols—mountains, rivers, trails, campsites, and more. Knowing these symbols can mean the difference between following a river to safety or accidentally heading up a cliff!

- **Contour Lines** – These wavy lines show elevation. The closer they are together, the steeper the terrain. If they're spread out, it means the land is flatter.

- **Compass Rose** – Somewhere on the map, you'll see a little compass showing N, E, S, W (North, East, South, West). These are the cardinal directions, and they're crucial for using a map with a compass.

- **Map Date & Magnetic Declination** – Maps can become outdated! Over time, magnetic north shifts slightly. The map will show the magnetic declination (how far magnetic north is from true north). If you're using a compass, knowing this difference helps keep you on track!

Understanding a map is great, but if you mix up north and south, you could be walking the wrong way for miles! That's why it's important to get your directions straight first.

Tricks for Remembering Directions

If you ever get mixed up with North, East, South, and West, try these fun memory tricks:

- "Never Eat Soggy Waffles"
- "Never Eat Silk Worms" (if you're feeling extra adventurous!)
- Make up your own saying!

Finding Where You Are on a Map

At first, figuring out where you are on a map might feel like trying to find a hidden treasure without an X marking the spot. But don't worry—it gets easier with practice!

- **Start with an Adult Guide** – If maps seem confusing, ask an adult to help. They can show you how to match real-life landmarks (like lakes, mountains, or roads) to what's on the map.
- **Practice While Traveling** – Next time you're in a car (and NOT the driver!), grab a road map. Follow along with your finger as you move, checking for towns, street names, or rivers. See if you can guess where you

are before a sign tells you!

- **Look Up!** – Don't get lost in the map! Every so often, look up and check your surroundings. Spotting landmarks like bridges, large trees, or buildings can help you confirm your location.

A great navigator hones (sharpens and improves) their map-reading skills before venturing into the wild. So grab a map, decode its secrets, and get ready for your next adventure!

Mastering the Compass: Navigate Like a Pro!

A compass isn't just a cool-looking gadget—it's a powerful tool that can guide you through the wild... if you know how to use it! But don't worry, you won't need a Ph.D. in navigation—just a little practice and a sense of adventure.

Meet Your Compass: The Parts That Matter

There are many types of compasses, but if you're just starting out, go for one with a clear baseplate—this makes it easier to plot your path on a map. Here's a quick tour of your new best friend:

- **Direction-of-Travel Arrow** – Always points

ahead—where you're going, not where you've been!

- **Rulers** – Found along the edges to help measure distances on your map.

- **Index Line** – The spot where you read your direction (bearing).

- **Rotating Bezel** – A numbered dial (0–360°) that helps set your course.

- **Magnetized Needle** – One end (usually red) always points north—like a tiny superhero that never gets lost.

- **Orienting Arrow** – A helpful outline inside the needle housing—line this up with the needle to stay on track.

- **Orienting Lines** – Grid-like lines that should match the map's gridlines for accuracy.

Some compasses even let you adjust for magnetic declination (the difference between magnetic north and true north). If yours doesn't, don't worry—you'll just need to do some quick math. Maps are a great start, but to really take control of your navigation, you need a compass. With a map in one hand and a compass in the other, you'll never be truly lost!

A Tale of Two Norths: The Secret to True Navigation

Did you know there are two kinds of North? It's true! Let's meet them:

- **True North** – This is the real North Pole, the top of the Earth. Every map is drawn using True North as the reference point.

- **Magnetic North** – This is where your compass points, but it doesn't stay in the same place! Earth's magnetic field is always shifting, so Magnetic North slowly moves over time.

Why Does This Matter?

Since Magnetic North shifts over time, you'll need to adjust your direction when using a compass with a map. This adjustment is called **magnetic declination.**

Maps will tell you the magnetic declination for that area. It's written as a number with E (east) or W (west).

How to Adjust for Magnetic Declination

If your map says Magnetic Declination: 12° East, here's what to do:

- If the declination is East, subtract it.
- If the declination is West, add it.

Example:

- If your bearing (the direction you're going) is 200° and the magnetic declination is 20° west, add 20° → New bearing: 220°
- If the declination is 20° east, subtract 20° → New bearing: 180°

How Often Should You Adjust?

Magnetic North moves a little bit every year, but maps don't change that fast! If you're using an older map, check when it was made. If it's from several years ago, the declination might be slightly different, so you may need to look up the most recent number.

Magnetic North isn't just important for hikers—it's used by GPS systems, pilots, the military, and even scientists to help with navigation! So when you adjust for declination, you're using the same skills as real explorers and professionals.

Now that you know the secret to navigating like a pro, let's put your skills to the test and get your compass and map working together!

Orienting Your Compass (So It's Actually Useful!)

Before you start trekking, you need to match your compass to your topographic map. Here's how:

1. Lay your map flat with north at the top
2. Place the compass on the map with the direction-of-travel arrow pointing north.
3. Slide the compass so the edge lines up with the map's

border.

4. Hold the baseplate still and twist the bezel until the N-symbol lines up with the index line.

5. Pick up both the map and compass, then turn your body until the magnetized needle is inside the orienting arrow.

Now your compass is aligned with the map! You'll need to do this each time you switch maps.

Taking a Bearing: Find Your Way Like an Explorer

A bearing is the direction you need to travel to reach your destination. Here's how to find yours:

1. **Mark Your Start and End Points** – On your map, mark where you are (Point A) and where you want to go (Point B).

2. **Draw a Line** – Use a ruler to connect the two points.

3. **Align Your Compass** – Place the corner of the compass on Point B, making sure the edge follows the line.

4. **Set the Bezel** – Rotate it until the orienting lines

match the gridlines on the map.

5. **Read the Bearing** – Check the number at the index line—this is your direction.

6. **Adjust for Declination** – If needed, adjust your number for magnetic declination.

7. **Follow the Arrow** – Hold the compass flat, turn your body until the magnetized needle is inside the orienting arrow, and look up. The object ahead of you is your target!

Now, start walking! But don't stare at your compass the whole time—pick a landmark in the distance and head toward it. Once you reach it, check your compass again, find another landmark, and repeat.

Compass Care: Keep It Working Like a Charm

A compass is an essential survival tool, so treat it with respect!

- **Don't drop it** – A damaged compass = bad directions.
- **Keep it away from magnets** – Magnets can mess with the needle's accuracy.
- **Read the instructions** – Every compass is a bit different, so get to know yours!
- **Wear it on a lanyard** – Avoid losing it in the middle of nowhere.

Mastering a compass takes practice, but once you get the hang of it, you'll never feel lost again. Whether you're hiking through dense forests, navigating the mountains, or just showing off your survival skills, this tiny tool is your key to adventure!

Navigating Without a Map or Compass

Even the best explorers can lose their map or forget their compass. But don't worry—nature is full of clues to help you find your way!

If You Get Lost, Remember:

- **DON'T panic.** Stop, take a deep breath, and think before you move.
- **DON'T wander aimlessly.** Walking in the wrong direction can make things worse.
- **DO look for landmarks.** A mountain, river, or even a unique tree can help you track where you are.
- **DO leave markers.** Stack rocks, break branches, or draw arrows in the dirt to mark your path.

But don't wait until you're lost to practice! Next time you're in a park or forest, try this. Walk a short distance, then turn around. Can you find your way back using landmarks?

Using the Sun to Find Direction

The sun is one of the easiest ways to figure out where you're going. Here's what to remember:

- The sun rises in the east and sets in the west.
- At midday (noon), the sun is in the south if you're in the Northern Hemisphere.
- If you're in the Southern Hemisphere, the sun is in the north at midday.

Try this: Stand outside at noon and notice where the sun is. Can you tell which way is north or south?

Using the Stars to Navigate

When the sun goes down, the stars can still guide you!

- In the Northern Hemisphere: Find the Big Dipper (it looks like a soup ladle). The two stars at the end of the Dipper's "bowl" point straight to the North Star, which tells you where north is.
- In the Southern Hemisphere: Look for the Southern Cross (a tilted cross-shaped constellation). Draw an

imaginary line down from the long part of the cross to the horizon—that's south!

Now, let's put your navigation skills to the test!

Activity 1: Find Your Way with the Shadow Stick Method

Even without a compass, the sun can help you figure out which way you're going! Your first challenge is to navigate using the Shadow Stick Method.

What You Need:

A straight stick (about a foot long)

Two small rocks or markers

A sunny spot

How to Do It:

1. Push the stick into the ground in an open area.

2. Mark the tip of its shadow with a rock.

3. Wait 30-60 minutes, then mark the new tip of the shad-

ow.

4. Draw a line between the two marks—this line runs west to east!

5. Stand with the first mark on your left, and you'll be facing north!

Bonus Challenge:

Try this at different times of the day. How does the shadow change? Does it move faster or slower at different times?

How Did You Do?

- **Explorer** – Tried it but couldn't find a shadow.
- **Navigator** – Found the shadow and figured out directions.
- **Master Navigator** – Tracked the shadow multiple times and used it to navigate!

Activity 2: Find North Using the Stars

If you're lost at night, don't worry—the stars have been guiding explorers for centuries! Your next challenge is to find your way by using the night sky.

What You Need:

A clear night sky

How to Do It:

In the Northern Hemisphere:

1. Find the Big Dipper (it looks like a giant soup ladle in the sky).

2. Look at the two stars at the end of the Dipper's "bowl."

3. Draw an imaginary line straight up. These two stars point directly to the North Star (Polaris), which is the tip of the handle of the Little Dipper!

4. Once you've found the North Star, you know exactly where north is!

In the Southern Hemisphere:

1. Look for the Southern Cross (a tilted, cross-shaped constellation).

2. Draw an imaginary line straight down from the long part of the cross to the horizon.

3. Where that line meets the ground—that's south!

Bonus Challenge:

Can you spot these constellations in your backyard? Try using a stargazing app or a star chart to help you identify them!

How did you do?:

- **Star Searcher:** No idea what you were looking at, but you gave it a try!

- **Star Gazer:** Found the constellation you were looking for.

- **Astronomer in the Making:** Spotted at least one constellation, found the North or South Star, and now you know which way you're going!

Activity 3: Walk a Triangle!

Navigating in the wilderness takes practice, and one of the best ways to sharpen your skills is by walking a perfect triangle using only your compass!

What You Need:

A compass

A marker for your starting point (rock, stick, whatever works)

A large, open area

How to Do It:

1. Set your compass to 0° (North) and walk 100 steps forward.

2. Turn the bezel to 120°, align your needle, and walk another 100 steps.

3. Turn the bezel to 240°, align your needle, and walk 100 more steps.

If you did it right, you should be right back where you started! If not, check that your starting bezel was at 0° and try again.

No Compass? No Worries!

Try walking a triangle by estimating your turns!

- Pick a starting point and walk 100 steps forward in a straight line.
- Turn your body about a third of a full turn (120°) and walk 100 more steps.
- Turn another 120° and walk your final 100 steps.

Did you end up close to where you started? If not, try again and adjust your turns!

Bonus Activity: Compass Scavenger Hunt

Want a little extra practice with your navigation skills? Create your very own Compass Scavenger Hunt and challenge your friends or family to follow your directions!

What You Need:

A compass

A notepad and pencil

Small objects to hide (like a pinecone, rock, or toy)

An open outdoor space

How to Do It:

- Set Up the Hunt: Hide 3-5 small objects around your area. For each item, write down directions using your compass (e.g., "Start at the big tree. Walk 20 steps north, then turn 90° east and walk 15 steps.")

- Create a Compass Rose: Draw a simple compass rose on your notepad with N, S, E, and W to help keep track of directions.

- Challenge a Friend: Hand them the first clue and let them follow your directions to find the hidden objects!

- Switch Roles! Once they've found the treasures, let them set up a scavenger hunt for you!

Now grab your map and compass (or keep practicing your natural navigation skills) and go chart your next expedition!

Wilderness Warrior Skill Unlocked!

You've mastered the skill of navigation! Now, mark your progress by coloring in your Wilderness Warrior Navigation Award on your Survivalist Skill Tracker (page 279).

Level Up Your Survival Skills!

Now you know how to find your way through the wild! But what if someone gets injured? That's why learning first aid is essential—it could make all the difference in an emergency.

Up next: Skill 8 – First Aid for Survival! Let's GO!

Survival Skill 8:
First Aid

"Do what you can, with what you have, where you are." — Theodore Roosevelt

You're climbing over a log when—ouch!—a branch scratches your leg. Or maybe you touch a really hot rock by the fire. Or worse—you trip and take a hard fall. What do you do?

Out in the wilderness, there's no doctor's office, no quick trip to the pharmacy, and no Band-Aids magically appearing. That means **YOU** have to be ready to handle small injuries—and stay calm in bigger emergencies.

Knowing first aid isn't just about patching yourself up—it's a survival skill that could save someone's life. Maybe even your own. Let's make sure you know what to do when things don't go as planned.

What Is First Aid?

First aid is the help someone gives right after an injury—before a doctor or rescue crew arrives. And guess what? You don't have to be a professional to help. Even kids can learn life-saving first aid skills!

There are three main levels of first aid (some kids even take official classes). The more you learn, the more you can do. But even basic skills—like stopping bleeding or helping someone stay calm—can make a huge difference.

Recognizing Serious Emergencies

In the wild, you might run into small problems like cuts, bruises, or bug bites. But some injuries are more serious. Here's how to tell when it's time to call for help right away:

- The person is bleeding a lot and it won't stop.
- They're unconscious (won't wake up or respond).
- They have trouble breathing.
- Their skin is burned badly or looks charred.

These are signs of a serious emergency—too big to handle alone. First aid can help stabilize the person, but you'll still need back-up fast.

Call–Check–Care: What to Do in an Emergency

When something serious happens, remember these three steps:

1. CALL:

Get help fast! Yell for an adult or use a phone to call 911 (or your local emergency number). If you're alone, find someone who can help.

2. CHECK:

Look at the person. Are they awake? Breathing? Bleeding? Is it safe for you to help?

If it's too dangerous (fire, falling rocks, wild animals), stay back and wait for help.

3. CARE:

Do what you can with what you know. That might mean:

- Helping someone sit or lie down
- Stopping bleeding with pressure

- Keeping someone calm and warm
- Using first aid skills you've practiced

Remember: You don't have to do everything—just doing something can make a big difference!

Emergency! When Every Second Counts

What if you face a situation that's really serious—and every second counts? Don't worry. Here are three of the most important things to know if someone is badly hurt:

1. Bleeding That Won't Stop

If someone is bleeding a lot, you need to act fast.

- **Apply pressure right away.** Use a clean cloth or bandage. Press down hard with your hands.
- **Don't lift the cloth to check it.** Keep pressing!
- If the bleeding soaks through, **add more cloth on top**—but don't remove the first one.

- If you can, **raise the injured area** above their heart.
- If you need your hands free, or if it's on an arm or leg, **you can tie fabric snugly around the bandage to keep pressure on.** (Not too tight! You still want blood to reach fingers or toes.)
- Keep holding pressure until help arrives.

Important: Only trained people should use a real tourniquet. For everyone else, steady pressure is the best way to help until help arrives.

2. If Someone Is Unconscious

If someone won't wake up, you need to check two things:

- **Are they breathing?** Put your hand on their chest to feel it rise and fall. Put your ear near their nose and mouth to listen and feel for air.
- **If they are breathing**, roll them gently onto their side—this is called the **recovery position**. It helps keep their airway clear while you wait for help.

Don't shake them or try to make them sit up. Keep them warm and stay close by.

3. If Someone Is Not Breathing

This is the most serious emergency.

- **Call or find help right away,** but don't leave the person alone. People can only live a few minutes without breathing.

- If you know CPR and you're trained to do it, begin right away.

- If the person is **choking**, they might grab their throat, cough, or not be able to make any sound. If this happens:

    - Ask if they're choking.

    - If they nod or can't talk, and you know how, **use the Heimlich maneuver** (abdominal thrusts) to help clear their airway.

    - If you don't know how—**get an adult fast!**

Pro Tip: CPR is a life-saving skill! If you want to learn it, ask an adult to help you sign up for a first aid or CPR class for kids.

How to Signal for Help

First aid means helping someone until more help arrives—but sometimes, you can't fix everything on your own. Even if you know how to make a splint (a sturdy support for a hurt arm or leg), that doesn't mean the person can walk out of the woods. That's why one of the most important survival skills is knowing how to signal for help—so rescuers can find you quickly when things go wrong.

No phone? No problem. Here are a few smart ways to get attention:

- **Bright colors** – Tie bright or colorful fabric or gear to trees where rescuers can see it.
- **Emergency whistle** – Blow 3 sharp blasts—the universal signal for help.
- **"HELP" sign** – Spell it out on the ground with sticks, rocks, or bright gear in an open space where it's easy to spot from above.

Once the injured person is safe and calm, it's time to make sure someone finds you.

First Aid = The 3 Ps

Every first aid situation has three big goals. Remember the **3 Ps**:

- **Preserve Life** – Keep the person alive.
- **Prevent More Harm** – Don't let things get worse.
- **Promote Recovery** – Help the person stay stable until more help arrives.

But before you can help others, you need to stay calm yourself. Let's talk about some tricks to do that.

How to Stay Calm in an Emergency

Your best survival tool isn't in your backpack—it's your head! But thinking clearly and acting quickly only works if you stay calm under pressure. Try these techniques:

1. Deep Breathing Method

- Breathe in slowly for 3 seconds.
- Hold your breath for 3 seconds.
- Breathe out slowly for 3 seconds.
- Pause for 3 seconds.
- Repeat until your heart rate slows and your mind clears.

2. The 10-Count Focus Trick

- Slowly count to 10 while naming 10 things around you.
  ("Tree. Rock. Sky. Shoe...")
- This helps your brain focus and keeps panic from taking over.

Just remember, the best kind of helper is a calm one!

Improvised First Aid Kit

It's smart to carry your own first aid kit. But what if it gets lost or left behind? Don't worry—there are lots of ways to treat injuries using supplies from your backpack or with what's around you.

Try these tools:

- **Extra clothing** – Use as a bandage, sling, or padding for a splint.

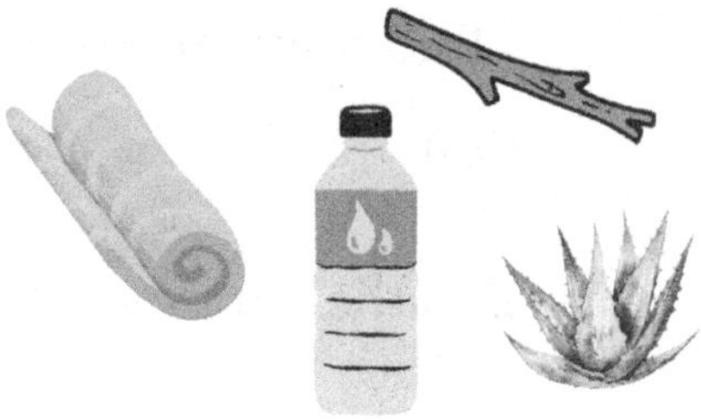

- **Sturdy sticks** – Great for making a splint or crutch.
- **Purified water** – If you don't have antiseptic, clean water helps flush wounds.
- **Aloe gel** – Found inside aloe leaves. Helps soothe burns or cuts. (Test it first—some people are allergic.)

Pro Tip: Believe it or not, honey is a natural antibiotic! It helps prevent infection and never spoils if kept clean. Ancient civilizations

used it to treat wounds, and you can, too. A small sealed container in your pack can come in handy.

Common Injuries and How to Treat Them

Out in the wild, bumps, scrapes, and surprise stings can happen to even the most careful adventurers. That's why it's important to know what to do—**before** you need to do it! While getting trained in first aid is the best option, learning the basics now means you'll be more prepared to help if something goes wrong.

First things first: Always wash your hands before helping someone who's hurt. If you have clean gloves, even better—put them on!

Scrapes and Cuts

Scrapes happen when the top layer of skin gets scratched or rubbed off. Cuts go deeper and may bleed more. Here's what to do:

- Press a clean cloth or gauze on the cut and hold it for 15–20 minutes to stop the bleeding.
- Rinse the wound gently with clean water.
- Check for dirt or pebbles. Use tweezers (wiped clean) to

remove anything stuck inside.

- Apply antiseptic cream, aloe, or honey.
- Bandage the cut. Scrapes can be left open unless they're still bleeding.

Get help if:

- The bleeding doesn't stop.
- The wound gets red, swollen, or has pus (a sign of infection).

Bruises

Bruises are blue or purple spots caused by blood vessels breaking under the skin. They might hurt, but they're usually not serious.

- Gently press a cold pack on the bruise for 20 minutes.
- Keep the bruised area raised above the heart to help reduce swelling.

Get help if:

- The bruise is very large or on the chest or stomach.
- A lump appears and doesn't go away, it might be a clot.

Burns

Burns can happen from hot objects, fire, or even too much sun. How serious a burn is depends on how deep it goes into the skin—this is called the *degree* of the burn.

- **First-degree burns** are the mildest. The skin turns red and feels sore, like a sunburn.
- **Second-degree burns** are more painful. They may have red skin, swelling, and blisters.
- **Third-degree burns** are very serious. The skin might look white, black, or burned. These burns go through all the layers of skin and need emergency help.

Only treat **first- and second-degree burns** yourself. **Get help right away** if the burn is worse.

How to Treat a Burn:

- Rinse the burned area with **cool (not cold!) water** for 10–15 minutes.
- Gently pat the area dry.
- Apply **aloe, burn cream, or honey** to soothe the skin.

- Cover it with a **clean bandage** or **plastic wrap** to keep it clean.

Never:

- Use ice (it can make the burn worse).
- Pop blisters (this can cause infection).
- Put tight clothing over a burn.

Get help if:

- The skin looks white, gray, or charred.
- The burn is **larger than 3 inches**.
- The burn is on the **face, hands, feet, or private areas**.

Nosebleeds

Got a nosebleed?

Don't tip your head back—that just sends blood into your stomach (yuck!).

- Sit up straight and lean forward.
- Gently blow your nose to clear out clots.
- Pinch the soft part of your nose and hold for 10–15 minutes.
- Breathe through your mouth while you wait.

Get help if:

- The bleeding doesn't stop after 30 minutes.
- You feel dizzy or it happened after a fall.

Splinters

Splinters are tiny bits of wood, metal, or thorns that get stuck in your skin. Most come out on their own—but some need help.

- Wash the area first.
- Use clean tweezers to gently pull it out.
- A pin (with adult help!) or sticky tape can help remove stubborn ones.
- Baking soda paste works too—apply and wait a few hours.

Get help if:

- The splinter is under a nail or very deep
- It's bleeding a lot or might need a tetanus shot.

Poison Ivy & Poison Oak

These itchy plants can ruin your day! Learn to recognize them:

- **Leaves of three? Let it be!**
- Poison ivy has shiny, pointed leaves on a vine.
- Poison oak looks like an oak leaf on a bush.

If you touch them:

- Wash your skin right away (within 10 minutes!)
- Carefully remove any clothes that may have brushed against it.

Treat it by:

- Soaking a cloth in cool water and pressing it on the rash.
- Using calamine lotion, hydrocortisone cream, baking soda paste, or oatmeal baths.

Get help if:

- You have trouble breathing or swallowing.
- It's very painful or spreading fast.

Stay Safe: How to Prevent Injuries

The best kind of injury is the one that never happens! Here's how to avoid common problems in the first place:

- **Wear the right gear**—good boots, dry socks (wool is better than cotton), and clothes that match the weather.
- **Always pack an emergency kit**—and know what's inside.
- **Watch where you step**—twisting an ankle on a root or rock is no fun!
- **Learn how to spot animal tracks** (called **spoor**) so you can avoid wild surprises.
- **Never wander off alone.**
- **Always tell someone where you're going** and when you'll be back.
- **Most of all—use common sense** and if something doesn't feel safe, don't do it.

Activity 1: Make a Makeshift Splint

If someone hurts a bone or joint, you can help keep it from getting worse by using a **splint**. A splint holds the injured part still so the bone doesn't move around—which could cause more damage or pain. Keeping it steady helps protect blood vessels and nerves while you wait for help.

Pretty awesome that you can do that, right?

What You Need:

- **Something to wrap with:** bandages, cloth strips, belts, neckties, or tape
- **Something sturdy:** popsicle sticks (for fingers), a rolled-up newspaper (for arms), or sticks (for legs)
- **Padding:** folded fabric, cotton, or even a sock
- **A partner** (because survival skills are more fun with friends!)

How to Make One:

1. **Pick roles** – One of you is the injured explorer, the other is the rescuer. Choose which body part is "hurt"

(finger, arm, leg, etc.).

2. **Add padding** – Gently place it around the injury to protect the skin.

3. **Position the splint** – Place your sturdy object along the injured area to keep it straight and still.

4. **Wrap it up** – Secure the splint in place with your wrapping material. Wrap snugly but not too tight—you don't want to cut off blood flow!

5. **Do the nail test** – Press on a nail until it turns white, then let go. It should turn pink again quickly. If it doesn't, loosen the wrap a little.

Pro Tip: When someone is hurt, helping them stay calm is just as important as the splint. Reassure them by saying, "I'm here to help—you're going to be okay."

When you're done, switch roles and try again. The more you practice, the more prepared you'll be if someone ever really needs your help!

Activity 2: Make a Paracord Survival Bracelet

Paracord—short for *parachute cord*—is super strong and super useful in the wild. You can use it to:

- Build a shelter
- Hang up wet clothes
- Replace a shoelace
- Fix a broken backpack strap
- Or even make a fishing line!

But carrying a long bundle of rope? Not so fun. That's where a **paracord bracelet** comes in. With just 10 feet of cord and a little time, you can weave it into a cool bracelet you'll always have with you—just in case.

What You'll Need:

- 10 feet of 550 paracord (holds up to 550 lbs!) *(No paracord? No problem! Use rope, string, yarn, or a shoelace for practice.)*
- A 3/8-inch buckle (the clip kind)
- Scissors or a sharp knife
- A lighter (you'll need an adult's help)

How to Make It:

1. **Measure your wrist** and add 1 inch. This will be the final length of your bracelet.
2. Fold the paracord in half so the ends touch. This fold is the **middle**.
3. Push the fold through one side of the buckle. You'll see a loop.
4. Pull the loose ends of the paracord through the loop. Pull tight. This is called a **cow hitch**—it attaches the cord to the buckle.

5. Unclip the buckle, then thread the loose ends through the other buckle piece.

6. Slide the buckle along the cord so the space between both buckle pieces matches your wrist measurement from step 1.

Now for the fun part—**weaving the cobra knot!** If it feels too difficult, take your time to read and re-read the directions. Once you get it, it will be smooth sailing or shall we say quick weaving from there. You can do it!

You should now see:

- 2 **center cords** (these stay still)
- 2 **working cords** (the outer ones—you'll use these to weave)

Start Weaving:

1. Take the **right cord** and lay it **over the center cords**, leaving a loop on the right side.

2. Take the **left cord**, lay it over the right cord, then behind the center cords, and finally up through the right loop.

3. Pull both working cords tight. That's your first cobra knot!

4. Now do the same thing in reverse:

    - Start with the **left cord** over the center cords.

    - Take the **right cord**, wrap it over the left cord, behind the center cords, and up through the loop on the left.

5. Tighten again.

6. **Repeat the pattern—right, then left, then right again—until you reach the second buckle.**
Remember to switch sides each time, or your bracelet might start to twist.
If it does—no worries! Just undo a few knots, straighten it out, and keep weaving.
Mistakes are part of learning, and every knot you make gets you one step closer to mastering it!

7. When you're done, **cut off the extra cord**, leaving about 1/8 inch at the end.

8. **Ask an adult to help you melt the ends of the cord with a lighter.** While the melted part is still warm (but

not too hot!), they can press it gently against the other ropes using the bottom of the lighter. This helps seal it in place—like glue—so it won't unravel.

You Did It!

Wilderness Power on Your Wrist!

You've woven more than a bracelet—you've made a survival tool that's ready when you need it. As you build your skills, try out new weaves like the King Cobra or Fishtail. Want to keep going? There are awesome step-by-step videos on YouTube (like the ones from Paracord Planet) to help you learn even more. The more you practice, the more prepared you'll be—one knot at a time.

How to Weave a Cobra Knot Bracelet

Step-by-Step

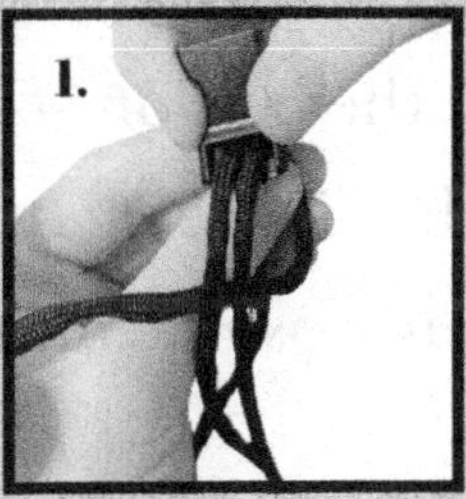

Start the Knot
Lay the right cord over the middle to make a loop.

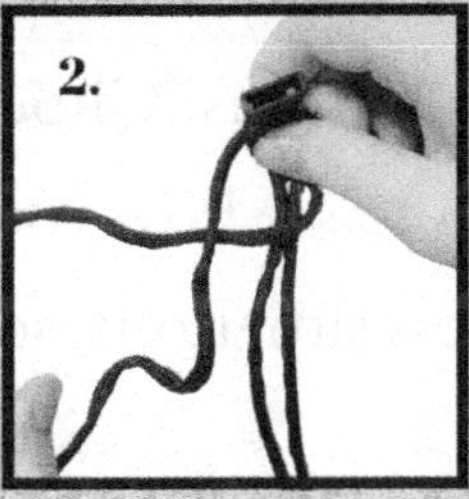

Cross Over
Bring the left cord over the right cord.

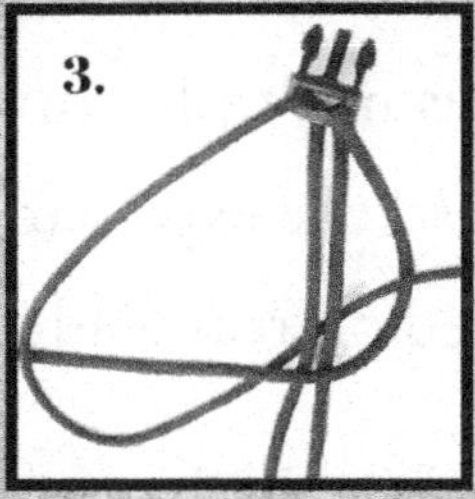

Under and Through
Wrap the left cord behind the center cords and pull it up through the right loop.

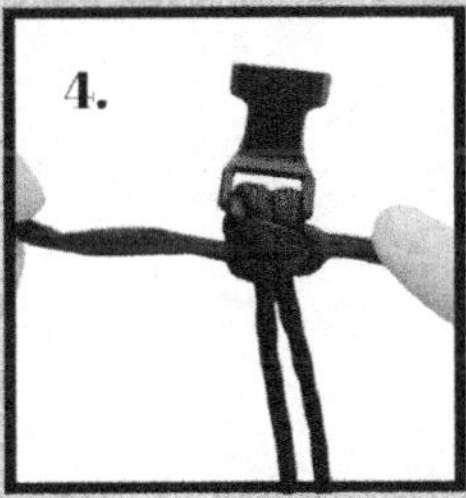

Tighten the Knot
Pull both cords tight to complete your first cobra knot!

Keep Weaving
Keep repeating the pattern–right, then left–until your bracelet is nearly finished.

Trim and Melt
Ask an adult to cut the ends and melt them carefully to keep the bracelet from unraveling.

Finished!

Awesome job! You've made your own cobra knot bracelet!

The cobra knot might seem tricky at first, but don't give up–you're going to master it in no time!

Activity 3: First Aid Scenarios – What Would You Do?

Let's put your first aid smarts to the test! Each situation below includes multiple-choice options. Choose the best one—and then read the explanation to see how you did. These questions are based on real wilderness situations, so pay close attention.

Ready? Let's go!

1. You scrape your knee while hiking. What's the first thing you should do?

A) Keep walking and let the air dry it out

B) Rinse it and cover if possible

C) Rub dirt in it—it helps it toughen up

Correct answer: B

If it's bleeding, rinse and cover *if you can*. Clean water helps flush out dirt and bacteria, but if you're low on water and the wound is small, it's okay to gently wipe it and keep it covered until you can clean it properly later. Never rub in dirt!

2. Your friend has a nosebleed. What should they do?

A) Lean forward and pinch the soft part of the nose

B) Lean back and stuff tissues up their nose

C) Lay down and try to breathe through their nose

Correct answer: A

Leaning forward helps prevent swallowing blood, and pinching the soft part helps stop the bleeding. Never lean back during a nosebleed—it can lead to nausea or choking.

3. You burn your hand while reaching near the fire. What's the safest way to cool the burn?

A) Put ice directly on it

B) Rinse it with cool (not cold) water for 10–15 minutes

C) Cover it with butter or oil—it'll soothe it

Correct answer: B

Cool water helps reduce damage and eases the pain. Ice can make a burn worse by damaging skin tissue. Butter or oil traps heat, which can also slow healing. Always use cool water and then apply something soothing like aloe if available.

4. Someone falls and hits their head hard. What's the first thing you check?

A) Ask if they're dizzy, confused, or feeling sick

B) Tell them to shake it off

C) Check for broken bones before anything else

Correct answer: A

If someone hits their head, always check for signs of a concussion: confusion, dizziness, headache, nausea, or trouble remembering things. Also, watch to see if they act "off."

Pro Tip: Apply something cool to reduce swelling, and have them sit down. If they lose consciousness, start vomiting, or can't focus, get help right away.

5. Your friend gets stung by a bee. What's the best thing to do first?

A) Use tweezers to pull the stinger out

B) Use a flat card to scrape it out

C) Tell them to ignore it unless it really hurts

Correct answer: B

Get an adult as soon as possible—especially if your friend has allergies or starts swelling up! If you need to remove the stinger, gently scrape it out with the edge of a card. Pulling with tweezers may squeeze more venom into the skin. Then apply something cool and watch carefully for allergic reactions (like trouble breathing or swelling).

6. Your friend twists their ankle on the trail. What should you do?

A) Wrap and elevate the ankle if possible, and let them rest

B) Walk it off—it'll feel better soon

C) Rub it and see if it loosens up

Correct answer: A

If possible, wrap the ankle and elevate it to reduce swelling.

Rest is best, but if you're still far from help, move slowly and carefully. Don't ignore the injury—moving too much can make it worse.

7. You find a splinter in your finger. What should you do first?

A) Wash your hands and the area with clean water

B) Try to yank it out quickly

C) Ignore it—it'll work its way out

Correct answer: A

Clean the skin before removing a splinter to prevent infection. If it's small, you can try tape or a baking soda paste. If it's deep or won't come out, ask an adult or leave it alone until you get help.

8. You brush against poison ivy while hiking. What should you do?

A) Scratch it right away

B) Be careful not to spread it and wash that area as soon as you can

C) Wrap it in cloth to stop the itch

Correct answer: B

The quicker you wash off the oils, the better. Use soap and water if possible—and be careful not to touch other parts of your body before washing your hands. Scratching spreads the rash and can lead to infection.

9. Your friend may have broken their arm. What should you do?

A) Try to straighten it out

B) Splint it and keep it still

C) Have them move it around gently

Correct answer: B

Never move or straighten a possible broken bone. Splint it to keep it stable and still, and get help as soon as possible.

10. You get a big bruise after falling. What helps most?

A) Splint it and wrap it tightly

B) Apply a cool pack and keep it raised

C) Poke it to check how deep it is

Correct answer: B

You don't need a splint for a bruise, but applying something cool and keeping it elevated can help reduce swelling. Avoid pressing or poking—it'll only make things worse.

How Did You Do?

First Aid Rookie (0–3 Correct): You're just getting started—review the tips and try again! You're already learning valuable skills.

Rescue Ready (4–7 Correct): You've got a solid grip on the basics. Just a little more practice and you'll be a first aid champ.

Life Saver Level (8–10 Correct): Awesome! You've got the calm, the smarts, and the know-how to handle injuries like a pro.

Wilderness Warrior Skill Unlocked!

You've mastered the skills needed for basic first aid! That means you're ready to stay calm, think clearly, and help when it matters most.

Mark your progress by coloring in your Wilderness Warrior First Aid Award on your Survivalist Skill Tracker (page 279).

Level Up Your Survival Skills!

Now you know how to handle injuries and stay safe! But let's be real—survival is tough, and a good meal can make all the difference.

It's time to learn how to cook like a survival pro.

Up next: Skill 9 – Campfire Cooking! Let's GO!

Survival Skill 9: Campfire Cooking

"Bananas are berries, but strawberries aren't. Botany is weird—and delicious!" — Fun Food Fact

Fuel for Your Adventure: Why Campfire Cooking Matters

Campfire cooking is more than just fun—it's fuel for your brain, your muscles, and your mission! Whether you're warming up after a chilly hike or powering up before a full day outdoors, knowing how to cook safely in the wild is one of the best survival skills you can learn.

But before you roast your first marshmallow or flip your first pancake over the fire, there's something even more important to talk about: food safety.

Stay Healthy Around the Campfire

Just because you're out in the woods doesn't mean you can skip good hygiene and safe food practices. In fact, getting sick from spoiled or contaminated food is one of the easiest ways to ruin your entire adventure. So here's what every Wilderness Warrior needs to know:

Keep Those Hands Clean!

Before you touch any food, wash your hands with soap and warm water—or use hand sanitizer if there's no sink nearby. This is especially important if you've been touching gear, handling raw meat, or playing in the dirt.

Watch the "Danger Zone"

Bacteria love the temperature range between **41°F and 140°F**—that's called the *danger zone*, and it's where germs multiply fast! Keep cold foods cold, and hot foods hot, so they stay safe to eat.

- Use a cooler packed with ice packs or frozen water bottles to keep meat, dairy, and salads below 41°F.
- Store raw meat at the bottom of your cooler so it doesn't drip on other food.
- Once cooked, keep food warm by placing it near the edge of the campfire on hot embers. Not too hot—you don't want it to dry out!

Separate Raw and Cooked Foods

Never use the same cutting board or utensils for raw and cooked food—this can spread dangerous germs and cause food poisoning. Keep your prep areas and tools separate!

Cook Food All the Way Through

Want to know if meat is ready? Don't just guess—use a **food thermometer** if you have one! Meat should reach an internal temperature of **167°F** to be safe. (Steak can be a little pink inside if the outside is cooked well—ask an adult if you're unsure.)

Don't Let Food Sit Out Too Long

If food is left out for **more than 4 hours**, bacteria can build up. If you're taking leftovers home, they should be packed away **within 2 hours**—and even sooner on hot days.

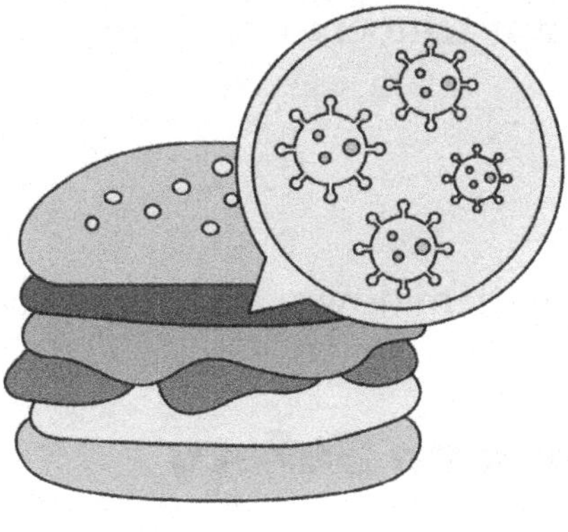

- Leftovers should be stored in clean, **sealed containers** after they've cooled.

- Keep food **covered at all times**—bugs like flies can carry germs, and you definitely don't want them on your dinner!

Remember to Keep It C.O.O.L!

4 Rules for Safe and Smart Campfire Cooking.

C – Clean:

Always start with clean hands and clean tools.

O – Organize:

Separate raw and cooked foods and use different utensils.

O – Overlook Nothing:

Don't overlook meat temperatures—use a thermometer when you can!

L – Leftovers:

Pack them up, cool them down, and keep them safe to eat later.

A Quick Word on Safety

Campfire cooking is awesome—but it also means **heat, flames, and sharp tools**. Be sure to ask an adult to help when you're using a knife, lighting a fire, or handling hot pans. It's not just safer—it's more fun when you're working together!

How to Cook Over a Campfire

There are lots of survival-style ways to make a meal over fire. Here are a few:

1. Stick Cooking

Roast marshmallows, hot dogs, or fish on a sturdy green stick (dry sticks can catch fire!). Hold above the coals and rotate often.

2. Rock Cooking

Heat a clean, flat rock near the fire and use it like a griddle—for eggs, bacon, or even bread!

3. Foil Packet Cooking

Wrap food in foil and place it in the coals. Turn often for even cooking—perfect for potatoes, veggies, or meat.

4. Campfire Grill or Pot

Use a metal grill or hang a pot to boil water or cook soup. Avoid placing cookware in direct flames—coals are better!

Pro Tip: No matter what you're cooking, don't cook in the flames—use the glowing coals instead! Flames are wild and uneven, but coals give steady heat and more control.

Now that you know how to cook safely and smartly outdoors, it's time to put your skills to the test!

Activity 1: Build a Solar Oven

If you needed to slow-cook your food in the wild, one option would be to wrap it in foil, place it on hot rocks from your firepit, cover it with more hot rocks, and then pile on dirt to trap the heat. Neat, huh?

But since we're not in the wilderness today, let's try a *solar oven* instead—powered by sunlight!

What You'll Need:

- A pizza box, shoebox, or any box with a hinged lid
- Aluminum foil
- Plastic wrap or clear sheet protector
- Black construction paper or cardstock
- Tape and glue stick
- Scissors
- Ruler
- Stick or straw (to hold flap open)

- Thermometer (optional, but fun!)

How to Build It:

1. Cut a Flap in the Lid:

Leave one side of the box (the hinged side) attached, and cut a square or rectangle shape in the lid on the other three sides. Leave about a 1-inch border around the cut edges. When you lift the flap, the remaining part of the lid will look like a TV screen or a frame.

2. Wrap the Flap in Foil:

Cover the inside of the flap with aluminum foil, making sure the shiny side is facing out. This helps reflect sunlight into the box. You can use glue, tape, or staples to hold the foil in place.

3. Make a Plastic Window:

Tape plastic wrap over the hole where you cut the flap. You can wrap it around the hole from both the inside and outside of the lid to double-layer it and trap more heat.

4. Line the Inside:

Glue black paper to the bottom of the box—this will help absorb heat. Then, cover the inside walls and the underside of the lid with aluminum foil. Make sure the foil is tight so it doesn't rip when you open and close the oven.

5. Angle the Flap Toward the Sun:

Find a sunny spot and adjust the flap so it reflects sunlight into the box. Use a stick or straw to prop the flap at the right angle.

6. Warm Up Your Oven:

Close the lid and leave the oven in the sun for 15–30 minutes. It should start to feel warm inside. If it doesn't, try moving the oven to a sunnier spot or try again at a different time of day.

7. Add Your Food:

Place your food, like s'mores, nachos, or mini pizzas, on a black tray or a foil-covered plate inside the oven. Then, close the lid.

8. Cook & Observe!:

Let the sun do the cooking! It may take 30–60 minutes. A good solar oven can get up to 150–200°F or even hotter. If your food isn't warm after 30–60 minutes, try finishing it in your home oven or wait for another sunny day to try again.

Pro Tip: The best time to use your solar oven is around midday when the sun is the brightest.

Activity 2: Plan a Two-Day Camp Menu

If you were going on a **two-night camping trip**, what would you eat? This activity challenges you to plan your meals and think through how to pack and prepare them.

What You'll Need:

- Paper and pencil
- Your imagination!

Your Challenge:

Plan out your meals for two days. Make sure to include one breakfast, one lunch, one dinner, and one dessert.

Bonus: Use recipes from the bonus section or create your own! Think about what ingredients you'll need, how you'll keep things hot or cold, and what gear you'll bring to cook it.

Pro Tip: Got a little extra salt, sugar, or flour?
If you're curious and have adult supervision, try these cool fire experiments next time you're around the campfire:

- A pinch of **salt** makes the flames burn a deeper orange
- A little **flour** creates a brief flare-up
- A sprinkle of **sugar** makes tiny sparks

Always ask an adult before adding anything to a fire—and never do it on a windy day!

How Did You Do?

- **Just Winging It** – Thought of a few meals.
- **Camp Planner** – Wrote meals down and considered packing.
- **PRO Planner** – Created a full list, included ingredients, gear, and color-coded fire needs!

Activity 3: Fire-Free Snack Hack

Not every survival snack needs a flame! When you pack smart, you can enjoy delicious food—no fire required.

What You'll Need:

- Access to a kitchen
- Ingredients like: nut butter, bananas, crackers, granola, cereal, trail mix, dried fruit, honey, oats, etc.

Your Challenge:

1. **Invent a No-Cook Camping Snack!**
   Try something like:
    - Apple slice sandwiches with nut butter and oats
    - DIY trail mix with cereal, seeds, and dried fruit
    - Graham cracker "campwiches" with banana and nut butter
2. **Name Your Snack:** What's it called?
3. **Write Down the Recipe:**

MY FIRE-FREE TRAIL TREAT

Name of Recipe:

Prep Time :

☆☆☆☆☆

Ingredients :

How to Make It:

Notes:

Wilderness Warrior Skill Unlocked!

You've cooked, planned, packed, and prepped like a true survivor.

Mark your progress by coloring in your **Campfire Cooking Award** on your Survivalist Skill Tracker (page 279).

Ready for your final test?

You've nearly completed your Wilderness Warrior journey. Flip the page and get ready for your **Final Wilderness Warrior Challenge**!

Final Wilderness Warrior Challenge:

Choose Your Survival Adventure!

You've learned how to build shelters, brave the elements, start fires, find water, forage and cook food—and even save a life.

Now it's time to prove just how much you've learned... by planning your own survival adventure!

Grab a pencil and some paper—it's time to see what you're made of.

Step 1: Choose Your Adventure!

Where are you headed?

- A hike through the deep woods
- A remote camping trip
- Exploring a wild, unfamiliar area
- Or... your own idea:

Got it? Great! Imagine you're really heading out tomorrow. Now let's pack.

Step 2: What's In Your Pack?

List 5–10 survival items you'd bring—and explain why you chose each one.
(Hint: Think shelter, food, safety, navigation, and first aid.)

Need more gear? Add it! Smart Wilderness Warriors are always prepared.

Step 3: Plot Twist! Something Goes W rong...

Your adventure is going great—until something unexpected happens.

Pick your challenge:

- Someone gets injured
- You lose your way
- A bear stole all your food
- The weather takes a wild turn
- All your water spilled!
- Or... your own wild twist:

Now what?

Step 4: Be the Hero — Solve the Problem!

Use your survival smarts to stay safe and overcome the challenge.
Here's the catch—you can only use what's in your pack and what you can find around you!

Ask yourself:

- Do you need to build a shelter?
- Signal for help or rescue?
- Help the injured person?
- Forage for food or purify water?
- Navigate using the sun, stars, or compass?

Write a few sentences (or more!) about how you solve the problem like a true Wilderness Warrior.

Step 5: Tell Your Tale!

Now it's time to bring your adventure to life. Choose how you'll tell your story:

- Draw a scene from your survival adventure
- Create a wilderness map with symbols
- Write and illustrate a short story
- Record the journey as a journal or diary
- Make a comic book with dialogue and action
- Or... something else creative:

When you finish, give your project a title—and don't forget to sign your name!

A Quick-Reference Refresher

Quick-Reference Survival Skills Checklist:

- **Safety:** Use STOP (Stop, Think, Observe, Plan).
- **Shelter:** Lean-to, A-frame, debris, tarp shelters.

- **Fire:** Safely build, maintain, and extinguish a fire.
- **Water:** Locate, collect, and purify safe drinking water.
- **Food:** Identify edible plants; know what to avoid.
- **First Aid:** Treat minor injuries and know when to seek immediate help.
- **Navigation:** Use landmarks, compass basics, and find your way safely.
- **Cooking:** Prepare safe, tasty meals over a campfire.

Review this checklist often—you're prepared and ready!

You Did It, Wilderness Warrior!

You've battled the wild, mastered each skill, and conquered the **Final Challenge**—WOW!

From finding food to building shelters, from staying safe to cooking over a fire, you've proven you have what it takes to survive and thrive in the great outdoors.

You've earned your title as a true Wilderness Warrior!

Grab a pen or marker and sign your award on page 280!

Conclusion

Congratulations, Wilderness Warrior!

You did it!

You've climbed every challenge, learned awesome survival skills, and proven you've got what it takes to face the wild—and anything life throws your way.

Built a shelter? (check)

Started a fire? (check)

Found food, stayed safe, and helped others? (double check!)

You've earned your title as a **real Wilderness Warrior**—but guess what?

This isn't the end. It's just the beginning...

Survival Skills are Life Skills

The things you've learned don't just belong in the woods. You can use them anywhere—

On a hike, at the campsite, during family adventures, or even in everyday situations like helping at home or shopping with your family.

- **Planning ahead?** That's a survival skill.
- **Solving problems?** Survival skill.
- **Being brave, helpful, and kind?** You guessed it—**survival skills!**

Keep Fueling the Fire

Wilderness Warriors never stop learning, growing, or exploring.

Wherever you go, remember to:

- **Be prepared** – Think ahead, pack smart.
- **Be respectful** – Take care of nature and the people around you.

- **Be brave** – You've got the skills and strength for the challenge ahead.

- **Pass it on** – Teach others what you've learned!

Adventure Is Everywhere

There are trails to hike, fish to catch, fires to build, and mysteries to solve.

But the biggest adventure of all?

Life.

And you, brave warrior, are ready for it.

Now go explore!

Be bold. Be prepared. Keep growing.

You're a Wilderness Warrior—and your next adventure is waiting!

Mastered
SAFETY
SKILLS
WILDERNESS WARRIOR

Mastered
SHELTER
BUILDING
WILDERNESS WARRIOR

Mastered
FIRE
STARTING
WILDERNESS WARRIOR

Mastered
WATER
SEEKING
WILDERNESS WARRIOR

Mastered
WEATHER
WATCHING
WILDERNESS WARRIOR

Mastered
FORAGING
FOR FOOD
WILDERNESS WARRIOR

Mastered
NAVIGATION
TACTICS
N
W
E
S
WILDERNESS WARRIOR

Mastered
FIRST AID
ESSENTIALS
WILDERNESS WARRIOR

Mastered
CAMPFIRE
COOKING
WILDERNESS WARRIOR

WILDERNESS WARRIOR

CERTIFICATE

OF ACHIEVEMENT

Congratulations,

____!

*You've completed all the challenges,
mastered essential survival skills, and
shown courage, creativity, and resilience
every step of the way.
You have officially earned the title of:
Wilderness Warrior
Whether you're navigating the wild,
building a shelter, starting a fire, or
helping someone in need—you're ready
for any adventure.*

Signed: ____________ *Date:* ____________

A. Hunter

A. HUNTER

Bonus:

Campfire Cooking Recipes

What's Cooking in This Chapter

- **Rise and Refuel** – Breakfasts by the fire
- **Midday Meals** – Lunches and quick bites
- **Fireside Feasts** – Warm, hearty dinners
- **Trail-Time Snacks** – Fuel for every adventure
- **Wilderness Treats & Sweets** – Sweet rewards by the fire
- **No Fire? No Problem!** – Quick, no-cook recipes

Tasty Meals and Helpful Tips

There's something unforgettable about sitting beside an open fire—the smell of smoke, the crackle of flames, the cozy warmth against the cool night air. It's relaxing, a little wild, and just dangerous enough to keep you on your toes. Fire deserves respect—and so does campfire cooking!

Roasting food over flames sounds fun, but it can turn frustrating (or even risky) if you're not prepared. The good news? You will be. This chapter is packed with simple, tasty recipes and helpful tips to make your outdoor cooking a breeze.

And if cooking over a fire isn't an option—no problem! You'll find some no-flame snack ideas here too.

Campfire Cooking Hacks & Smart Packing Tips

Cooking outdoors is awesome—but it can be tricky to pack everything you want and need. That's why we've pulled together this list to make your packing and cooking a little easier!

Prep Like a Pro (Before You Go)

- **Chop & Portion at Home**: Cut up veggies, meats, and spices in advance. Store in resealable bags or small containers.

- **Freeze Your Meals**: Freeze chili, stew, or soup in flat bags—they double as ice packs and are easy to reheat.

- **Use Condiment Packets**: Bring ketchup, mustard, soy sauce, or mayo from takeout orders to save space and avoid leaks.

- **Dehydrate Your Own Food**: Great for longer trips! Just add boiling water to rehydrate.

Campfire Cooking Tricks

- **Foil Packet Power**: Wrap meat, veggies, seasoning, and butter in foil. Cook in the coals for a fast, delicious meal.

- **Orange Peel Muffins or Brownies**: Scoop out an orange, fill it with batter, wrap in foil, and bake in the coals. Fun and citrusy!

- **Flat Rock Griddle**: Clean a flat rock or bring a small cast iron skillet for eggs, pancakes, or bacon—no grill required.

- **Muffin Tin Magic**: Cook eggs, mini pies, or snacks right over the fire in a muffin tin.
- **Dough on a Stick**: Wrap crescent or biscuit dough around a clean stick and roast. Fill with jam or Nutella after. Yum!
- **Use Pinecones**: Pinecones burn hot and fast—perfect for getting water to boil quickly.

Chilling & Storing Smart

- **Freeze Everything**: Meats, sauces, water bottles—frozen food keeps your cooler cold longer.
- **Vacuum-Seal for Freshness**: No leaks, no mess, and a longer shelf life.
- **DIY Cooler Organization**: Use bags or bins to separate meals by day.
- **Hanging Food Bag or Bear Canister**: Keep critters out and your food safe. Use odor-proof bags if you're in bear country.

Clever Extras

- **Pill Organizer Spice Rack**: These tiny containers make a great way to pack spices!
- **Grill Grate or Cooling Rack**: Set it on rocks for a DIY cooking surface.
- **Reusable Bags & Silicone Cups**: Easy to wash and great for snacks or leftovers.
- **Headlamp for Night Cooking**: Hands-free and well-lit makes everything easier after dark.

A Few Important Reminders

- Fires are **hot**—use caution and ask for adult help when needed.
- Never leave your fire unattended—**forest fires are real**.
- Clean up after yourself—**pack it in, pack it out**.
- Animals like bears, squirrels, and raccoons love your cooking too! Store food securely and out of reach.

How to Make a Foil Packet for Campfire Cooking

You'll notice that lots of the recipes in this chapter use foil packets—because they're easy, customizable, and require *almost* no cleanup. Here's how to make your own:

What You'll Need:

- Heavy-duty aluminum foil
- Cooking oil or spray (optional, but helpful!)

Step-by-Step Foil Packet Instructions:

1. **Tear a Big Sheet:**
   Cut a piece of foil about 12–18 inches long (or longer for bigger meals).

2. **Add Your Food:**
   Place your ingredients in the center of the foil.

3. **Seal the Packet:**

   - Bring the long sides of the foil together and fold them

down a few times to seal the top.

  - Then fold up the ends like a burrito and pinch them tight.

  - Make sure it's sealed well—no leaks!

4. **Create a Tent Shape (Optional):**
   Leave a little space inside for steam to move around. This helps cook food evenly.

Cheese and Sticky Stuff Tips:

- **Oil the Foil:** Lightly grease the foil before adding food to prevent sticking.

- **Keep Cheese in the Middle:** Sandwich cheese between other ingredients so it doesn't melt onto the foil.

- **For Cheesy Toppings (like pizza):** Keep the cheese below the foil seam—or leave the top slightly open to avoid a sticky mess.

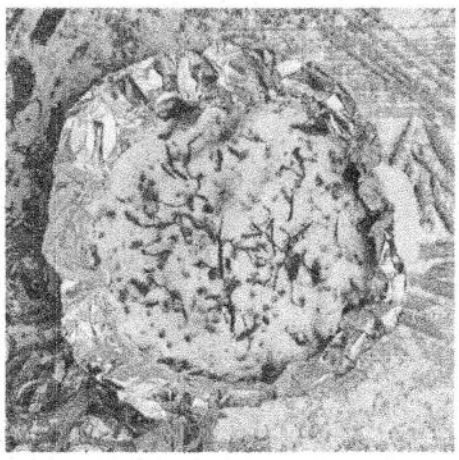

Flip with Care:

If your recipe allows it (like veggies or meat packs), flip the foil packet halfway through cooking using tongs. Just be gentle—no tearing!

Quick Notes Before You Begin

No Aluminum Foil? No Problem!

Most foil recipes can also be made in:

- A fire-safe pan with a lid
- A cast iron Dutch oven
- Over a grill grate with a little creativity!

Recipe Size:

These recipes are written to serve 1–2 people. So here's your math mission—multiply the ingredients by the number of hungry campers in your group!

Rate Your Recipes!

After each recipe, you'll find five empty stars for you to fill-in.

□□□□□ = Not my favorite

□□□□□ = Let's make this again!

Got your gear?

Got your fire?

Great—let's get cooking!

Rise and Refuel: Breakfasts

Eggs in a Nest

A fun way to cook an egg right inside your toast!

You'll Need:

- 1 slice of bread
- 1 egg
- Butter or oil
- Salt and pepper

How to Make It:

1. Use a cup to cut a circle in the center of the bread.
2. Butter a skillet and place it over hot coals.
3. Put the bread on the skillet and crack the egg into the hole.
4. Cook until the egg white sets on the bottom.

5. Flip carefully with a spatula and cook another minute.

6. Sprinkle with salt and pepper. Eat warm!

☆☆☆☆☆

Bacon on a Stick

Crispy bacon cooked over the fire, the wilderness way.

You'll Need:

- Thick-cut bacon strips
- A long, sturdy stick

How to Make It:

1. Wrap bacon around the stick like a spiral.

2. Hold it over the hot coals, not in the flames.

3. Turn slowly and cook until crispy, about 5–8 minutes.

☆☆☆☆☆

Bannock (Stick Bread)

Make old-fashioned survival bread over the fire!

You'll Need:

- 1 cup flour
- ½ tsp salt
- 1 tsp baking powder
- ¼ cup water
- 1 tbsp oil or melted butter

How to Make It:

1. Mix everything into a soft dough.
2. Roll into a rope and twist it around a clean stick.
3. Roast over coals, turning slowly until golden and cooked through.

☆☆☆☆☆

Campfire Oatmeal

Warm and hearty with your favorite mix-ins.

You'll Need:

- ½ cup quick oats
- 1 tbsp dried fruit
- 1 tsp brown sugar or honey
- 1 tbsp nuts or seeds
- Hot water

How to Make It:

1. Add oats, fruit, sugar, and nuts to a bowl or jar.
2. Pour in hot water and stir.
3. Let sit for 3–5 minutes, then stir and eat.

☆☆☆☆☆

Sunrise Quesadillas

Cheesy and warm with endless filling options.

You'll Need:

- 1 tortilla
- ¼ cup shredded cheese
- Optional: scrambled egg or beans

How to Make It:

1. Sprinkle cheese and filling on one half of the tortilla.
2. Fold it in half.
3. Cook in foil or on a skillet over the fire.
4. Flip once until golden and melty.

☆☆☆☆☆

Skillet French Toast

A sweet and cozy camp breakfast.

You'll Need:

- 2 slices bread
- 1 egg
- ¼ cup milk
- Butter
- Cinnamon
- Syrup

How to Make It:

1. Whisk egg, cinnamon, and milk together in a bowl or shallow dish.
2. Dip bread into the mixture.
3. Butter a skillet and place over fire.
4. Cook the egg-soaked bread on both sides until golden brown.

5. Drizzle the tops with syrup and enjoy!

☆☆☆☆☆

Midday Meals: Lunches & Quick Bites

Pita Pocket Lunch

An easy, handheld meal.

You'll Need:

- Pita bread
- Hummus, cheese, or deli meat
- Chopped veggies

How to Make It:

1. Fill pita with your favorite toppings.
2. Wrap in foil and warm near the fire if desired.

☆☆☆☆☆

Campfire Grilled Cheese

A classic sandwich, campfire-style.

You'll Need:

- 2 slices bread
- Cheese
- Butter

How to Make It:

1. Butter the outside of both bread slices.
2. Place the cheese between the slices.
3. Wrap in foil or cook in a skillet over the fire.
4. Flip until bread is golden and cheese is melted.

Cowboy Casserole Packet

A hearty lunch in a foil pouch.

You'll Need:

- Cooked ground beef
- Diced potatoes
- Cheese
- Corn or beans
- Seasoning

How to Make It:

1. Mix ingredients in foil.
2. Wrap tightly and cook in hot coals for 15–20 minutes.

Turkey & Cheese Roll-Ups

A simple wrap or no-fire option.

You'll Need:

- Sliced turkey
- Cheese sticks
- Tortilla or lettuce leaves

How to Make It:

1. Wrap cheese with turkey in a tortilla or lettuce.
2. Eat cold or warm briefly in foil.

Campfire Pizza Wraps

Pizza made easy, anywhere!

You'll Need:

- Tortilla
- Pizza sauce
- Cheese
- Toppings of choice

How to Make It:

1. Spread sauce on tortilla.
2. Add cheese and toppings.
3. Roll up and wrap in foil.
4. Warm near fire until cheese melts.

Fireside Feasts: Dinners by the Flame

Campfire Stew

A warm and hearty dinner by the fire.

You'll Need:

- Canned beans (any kind)
- Chopped vegetables
- Tomato sauce or broth
- Seasoning of choice

How to Make It:

1. Mix ingredients in a pot (do not use foil for stew).
2. Simmer over the fire or hot coals for 30–60 minutes.

Camp Baked Potatoes

Soft on the inside, crispy on the outside!

You'll Need:

- 1 potato
- Butter or oil
- Salt

How to Make It:

1. Poke a few holes in the potato with a fork.
2. Rub with butter or oil and sprinkle with salt.
3. Wrap tightly in foil and place in coals.
4. Cook for 30–60 minutes, turning halfway through.

Foil Packet Potatoes & Cheese

Simple and satisfying.

You'll Need:

- 1 potato, sliced
- ¼ cup shredded cheese
- Butter or oil
- Salt and pepper

How to Make It:

1. Place ingredients in the center of a foil sheet.
2. Wrap tightly and place in coals.
3. Cook for 15–20 minutes, turning once.

Camp Mac & Cheese

Cheesy and comforting!

You'll Need:

- 1 cup cooked macaroni
- ½ cup shredded cheese
- Splash of milk or cream

How to Make It:

1. Mix everything in a foil packet or small skillet.
2. Heat near coals until warm and gooey.

☆☆☆☆☆

Campfire Veggie Packets

A healthy side or full meal.

You'll Need:

- Mixed chopped vegetables
- Olive oil
- Seasoning

How to Make It:

1. Toss veggies with oil and seasoning.
2. Wrap in foil and seal well.
3. Cook in hot coals for 15–20 minutes, turning once.

Trail-Time Snacks: Fuel for the Adventure

Apple Slice Sandwiches

Crunchy, sweet, and easy to make.

You'll Need:

- Apple slices
- Nut butter or seed butter
- Oats, granola, or raisins

How to Make It:

1. Spread nut butter on one apple slice.
2. Sprinkle with oats, granola, or raisins.
3. Press another apple slice on top like a sandwich.

☆☆☆☆☆

Energy Balls for the Trail

No-bake bites full of energy!

You'll Need:

- ¼ cup oats
- 1 tbsp nut butter
- 1 tsp honey
- Dash of cinnamon or cocoa powder

How to Make It:

1. Mix all ingredients in a bowl.
2. Roll into small balls.
3. Eat right away or chill to take along.

DIY Trail Mix

Build your own mix just the way you like it.

You'll Need:

(pick your favorites or add other ingredients)

- Nuts
- Dried fruit
- Pretzels
- Seeds
- Chocolate chips

How to Make It:

1. Customize and mix everything together in a bowl or bag.
2. Pack in a container for your hike or snack time.

Fire-Roasted Nachos

A campfire twist on a crunchy snack.

You'll Need:

- Tortilla chips
- Shredded cheese
- Black beans
- Salsa or chopped veggies

How to Make It:

1. Layer chips, cheese, beans, and toppings in foil or a skillet.
2. Cover and place near coals.
3. Heat until cheese is melted.

☆☆☆☆☆

Corn on the Flame

A smoky and sweet treat.

You'll Need:

- Corn on the cob
- 1 tbsp butter
- ½ tsp garlic (optional)

How to Make It:

1. Rub corn with butter and garlic.
2. Wrap in foil.
3. Grill over coals, turning often until tender.

Pigs in a Sleeping Bag

Fun and filling!

You'll Need:

- Hot dogs
- Crescent or bun dough

How to Make It:

1. Wrap crescent dough around each hot dog.
2. Cook on a skewer or grill rack over the fire.
3. Turn until dough is golden and cooked through.

☆☆☆☆☆

Campfire Popcorn

Popcorn over the fire—so fun to watch!

You'll Need:

- 2 tbsp popcorn kernels
- 1 tsp oil
- Heavy-duty foil

How to Make It:

1. Tear off a large square of foil (at least 12x12 inches) and add kernels and oil in the center.
2. Fold the edges to make a roomy pouch, leaving space for the popcorn to expand. Seal tightly.
3. Attach the pouch to a stick or place gently on hot coals and shake gently.
4. Pop until the popping slows down. Remove carefully and let cool before opening.

☆☆☆☆☆

Campfire Queso Dip

A cheesy dip with a kick.

You'll Need:

- ¼ cup shredded cheese (cheddar or Mexican blend)
- 2 tbsp salsa
- Optional: black beans or jalapeños

How to Make It:

1. Mix ingredients in foil or a small pan.
2. Heat until melted.
3. Dip with chips and enjoy!

☆☆☆☆☆

Wilderness Treats & Sweets: Campfire Desserts

Roasted Cinnamon Apples

Warm, gooey, and full of flavor.

You'll Need:

- 1 apple
- Brown sugar or maple syrup
- Cinnamon
- Butter

How to Make It:

1. If you have an apple corer, core the apple and place it on a sheet of foil.
2. If not, you can slice the apple in half and scoop out a bit of the center with a spoon.
3. Add a little butter, a pinch of cinnamon, and a spoonful of brown sugar or maple syrup.

4. Wrap tightly in foil.
5. Place in hot coals and cook for 10–15 minutes until soft.

☆☆☆☆☆

Banana Boats

A melty mix of banana, chocolate, and marshmallows.

You'll Need:

- 1 banana
- Chocolate chips
- Mini marshmallows
- Crushed graham crackers

How to Make It:

1. Slice the banana down the middle without cutting all the way through. Leave the peel on.

2. Open slightly and fill with chocolate chips, marshmallows, and graham cracker crumbs.

3. Wrap in foil and heat in coals for 5–10 minutes.

Feel free to mix it up! Add toppings you love like chopped nuts for crunch, a spoonful of peanut butter, or a drizzle of caramel.

You can even skip the marshmallows or chocolate chips if they're not your thing.

Hey, whatever floats your boat!

☆☆☆☆☆

S'mores Cones

All the flavors of s'mores in a neat little cone.

You'll Need:

- Ice cream cone
- Marshmallows
- Chocolate chips
- Graham cracker crumbs

How to Make It:

1. Fill cone with marshmallows, chocolate chips, and graham crumbs.
2. Wrap in foil and place near coals for 3–5 minutes until melty.

Homemade Hot Chocolate

Warm up after a long day.

You'll Need:

- 2 tbsp cocoa powder
- 1 cup milk or water
- 2 tbsp sweetener (sugar, honey, maple syrup)
- Optional: cinnamon or vanilla

How to Make It:

1. Stir cocoa and sweetener together.
2. Mix into warm milk or water.
3. Stir until smooth and steaming.

☆☆☆☆☆

Grilled Pineapple Rings

Sweet and smoky tropical goodness.

You'll Need:

- Pineapple rings (slices)
- Honey or brown sugar

How to Make It:

1. Brush pineapple with honey or sprinkle with brown sugar.
2. Grill on foil or skewer over fire until caramelized.

☆☆☆☆☆

S'more-Stuffed Strawberries

A fruity twist on a classic treat.

You'll Need:

- Large strawberries
- Chocolate chips
- Mini marshmallows

How to Make It:

1. Cut out the center of each strawberry.
2. Fill with chocolate and marshmallow.
3. Wrap in foil and heat near coals for 3–5 minutes.

Cinnamon Sugar Donuts

Crispy and sweet with a soft inside.

You'll Need:

- Biscuit dough rounds
- Sugar
- Cinnamon
- Oil

How to Make It:

1. Heat oil in a pan over the fire.
2. Fry biscuit dough until golden brown.
3. Roll in a mix of cinnamon and sugar.

☆☆☆☆☆

Homemade Marshmallows

Make your own marshmallows from scratch!

Pro Tip: These should be made ahead of time, at least 1–7 days before your trip. Let them sit out in the open air after cutting so they firm up and travel well.

You'll Need:

- 1½ cups honey
- 3 tbsp gelatin
- 1 cup water, divided
- ¼ tsp salt
- 1 tsp vanilla
- Powdered sugar or arrowroot for dusting

How to Make It:

1. In a bowl, add ½ cup water and sprinkle in gelatin to bloom.

2. In a saucepan, boil honey and the other ½ cup water until it reaches 240°F.

3. Slowly pour the hot honey into the gelatin while mixing on low.

4. Whip on high speed until fluffy (about 10 minutes).

5. Add salt and vanilla.

6. Pour into a greased pan dusted with powdered sugar.

7. Let set for 4 hours, then cut and dust each piece.

☆☆☆☆☆

No Fire? No Problem!

You may have seen these recipes earlier in the book—but here they are all in one convenient place. These are perfect for rainy days, hikes, or when you just want something quick without cooking.

DIY Trail Mix

Build your own mix just the way you like it.

You'll Need:

(pick your favorites or add other ingredients)

- Nuts
- Dried fruit
- Pretzels
- Seeds
- Chocolate chips

How to Make It:

1. Customize and mix everything together in a bowl or

bag.

2. Pack in a container for your hike or snack time.

Apple Slice Sandwiches

Crunchy, sweet, and easy to make.

You'll Need:

- Apple slices
- Nut butter or seed butter
- Oats, granola, or raisins

How to Make It:

1. Spread nut butter on one apple slice.
2. Sprinkle with oats, granola, or raisins.
3. Press another apple slice on top like a sandwich.

Energy Balls for the Trail

No-bake bites full of energy!

You'll Need:

- ¼ cup oats
- 1 tbsp nut butter
- 1 tsp honey
- Dash of cinnamon or cocoa powder

How to Make It:

1. Mix all ingredients in a bowl.
2. Roll into small balls.
3. Eat right away or chill to take along.

Turkey & Cheese Roll-Ups

A simple wrap or no-fire option.

You'll Need:

- Sliced turkey
- Cheese sticks
- Tortilla or lettuce leaves

How to Make It:

1. Wrap cheese with turkey in a tortilla or lettuce.

2. Eat cold or warm briefly in foil.

Pita Pocket Lunch

Great for lunch without needing heat.

You'll Need:

- Pita bread
- Hummus, cheese, or deli meat
- Chopped veggies

How to Make It:

1. Fill pita with your favorite toppings.

2. Wrap and store in a cooler or eat fresh.

These recipes are perfect for rainy days, hikes, or anytime you can't use a fire!

And that's a wrap—Wilderness Chef!

Don't forget to color in the stars on the recipes you've tried. I hope you enjoyed a recipe or two and had a blast creating it! **Now go forth and feast, fearless Wilderness Warriors!**

Now, It's Your Turn!

Next, you'll find empty pages to write down your own wild recipes. What did you make? Who helped? Would you change anything next time? Don't forget to rate your masterpiece! *Use the next pages to write, sketch, or share your own wild creations.*

My Wild Recipe

Name of Recipe:

Prep Time :

☆☆☆☆☆

Ingredients :

How to Make It:

Notes:

My Wild Recipe

Name of Recipe:

Prep Time :

☆☆☆☆☆

Ingredients :

How to Make It:

Notes:

My Wild Recipe

Name of Recipe:

Prep Time :

☆☆☆☆☆

Ingredients :

How to Make It:

Notes:

My Wild Recipe

Name of Recipe:

Prep Time :

☆☆☆☆☆

Ingredients :

How to Make It:

Notes:

My Wild Recipe

Name of Recipe:

Prep Time :

☆☆☆☆☆

Ingredients :

How to Make It:

Notes:

Resources and Inspirations

The following resources provided helpful inspiration, facts, or guidance during the creation of *Wilderness Warriors: Survival Skills for Kids*. While not all material is quoted directly, these articles, guides, and websites helped inform content, tips, and project ideas throughout the book.

Stanley. (2020, November 12). *10 outdoor safety tips to protect yourself, others, & the environment*. https://www.stanley1913.com/blogs/how-to-guides/how-to-stay-safe-outdoors-and-recreate-responsibly

Bushcraft Buddy. (n.d.). *26 random survival facts*. https://bushcraftbuddy.com/26-random-survival-facts/

Adamant, A. (2019, October 2). *50+ edible wild berries & fruits ~ A forager's guide*. Practical Self Reliance. https://practicalselfreliance.com/edible-wild-berries-fruits/

Adina. (2024, July 3). *Wild garlic butter*. Where Is My Spoon. https://whereismyspoon.co/wild-garlic-butter/

Amy. (2020, July 9). *Naturally sweetened homemade marshmallows*. Healthy Happy Farm. https://healthyhappyfarm.com/honey-homemade-marshmallows/

Amy. (2024, April 3). *Cozy cocoa: Healthier homemade hot chocolate*. Healthy Happy Farm. https://healthyhappyfarm.com/cozy-cocoa-delight-homemade-hot-chocolate/

Anderberg, J. (2023, July 2). *How to find water in the wild*. The Art of Manliness. https://www.artofmanliness.com/skills/outdoor-survival/how-to-find-water-in-the-wild/

April. (2023, January 22). *10 fun weather activities for elementary students*. Grade School Giggles. https://www.gradeschoolgiggles.com/10-fun-weather-activities-for-elementary-students/

Ashley. (2021, May 15). *Camping hot dogs recipe for the campfire*. Frugal Coupon Living. https://www.frugalcouponliving.com/camping-hot-dog-recipe/

Barrow, C. (2019). *7 lightning safety tips if you're caught outside during a thunderstorm*. AccuWeather. https://www.accuweather.com/en/weather-news/7-lightning

-safety-tips-if-youre-caught-outside-during-a-thunderstorm-2/434483

Bauer, E. (2024, February 25). *Nettle soup*. Simply Recipes. https://www.simplyrecipes.com/recipes/nettle_soup/

Benjamin, C. (2023, April 24). *How to make a splint*. MedlinePlus. https://medlineplus.gov/ency/article/000040.htm

Bergo, A. (2024, May 1). *Classic stinging nettle soup*. Forager | Chef. https://foragerchef.com/classic-nettle-soup/

Bond, S. (2024, June 19). *Campfire mac and cheese (easy to prep!)* Live Eat Learn. https://www.liveeatlearn.com/campfire-mac-cheese/

Boshoff, W. (2017, October 31). *Lightning safety*. Hiking South Africa. https://www.hikingsouthafrica.co.za/lightning-safety/

Bushcraft Buddy. (n.d.). *5 things to keep in mind when choosing a shelter location*. https://bushcraftbuddy.com/5-things-to-keep-in-mind-when-choosing-a-shelter-location/

Campfire Safety. (n.d.). *Smokey Bear*. https://smokeybear.com/en/prevention-how-tos/campfire-safety

Camping mac and cheese. (2018, August 24). *Lauren's Latest*. https://laurenslatest.com/camping-mac-n-cheese/

Celley, C. (n.d.). *Don't touch these plants! Six lookalikes you want to avoid*. U.S. Fish & Wildlife Service. https://www.fws.gov/story/dont-touch-these-plants

Cloud types. (n.d.). *UCAR Center for Science Education*. https://scied.ucar.edu/learning-zone/clouds/cloud-types

Cunningham, K. (2023, November 10). *How to purify water in the wild: 6 methods to avoid an upset tummy in the backcountry*. Advnture. https://www.advnture.com/how-to/purify-water-in-the-wild

Desmazery, B. (2015, April). *Wild garlic butter*. BBC Good Food. https://www.bbcgoodfood.com/recipes/wild-garlic-butter

DIY Projects. (2018, May 17). *Learn how to make fire starters out of egg cartons*. https://diyprojects.com/make-fire-starter-egg-carton-dryer-lint/

Earhart, A. (n.d.). *Amelia Earhart quotes*. BrainyQuote. https://www.brainyquote.com/quotes/amelia_earhart_130007

Elges, N. (2023, December 11). *11 deadly plants you should never eat*. Backpacker. https://www.backpacker.com/skills/identify/plants/11-deadly-plants-you-should-never-eat/

Fowler, P. (n.d.). *How to treat poison ivy, oak, and sumac.* WebMD. https://www.webmd.com/allergies/understanding-poison-ivy-oak-sumac-treatment

Fresh Off The Grid. (2020, May 6). *Campfire banana boats 9 ways.* https://www.freshoffthegrid.com/campfire-banana-boats/

Fresh Off The Grid. (2021, June 10). *The easiest campfire nachos.* https://www.freshoffthegrid.com/campfire-nachos/

Furst, J. (2020, June 17). *How to build a first aid kit from household items.* First Aid for Free. https://www.firstaidforfree.com/how-to-build-a-first-aid-kit-from-household-items/

Grylls, B. (n.d.). *Bear Grylls quotes.* BrainyQuote. https://www.brainyquote.com/quotes/bear_grylls_512988

Hansen, D. (2024, July 18). *Campfire grilled nachos.* Let's Camp S'more. https://letscampsmore.com/grilled-nachos/

Hitchcock, K. (2024, April 24). *13 bushcraft and survival shelters – the ultimate guide.* Survival Stoic. https://survivalstoic.com/bushcraft-and-survival-shelters/

Jacobson, C. (2012, August). *An expert's tips for avoiding accidents in the wilderness.* Scouting Magazine. https://scoutingmagazine.org/2012/08/an-experts-tips-for-avoiding-accidents-in-the-wilderness/

James, H. (2022, December 20). *Wilderness navigation: The ultimate survival skill*. Unified Survival. https://unifiedsurvival.com/wilderness-navigation-the-ultimate-survival-skill/

Lirette, D. (2019, August 24). *"Hug the tree and wait": Search and rescue teaches kids what to do if they get lost in the woods*. C B C . https://www.cbc.ca/news/canada/british-columbia/search-and-rescue-teach-kids-what-to-do-lost-in-woods-1.5257078

Mayo Clinic Staff. (2024, April 17). *Bruise: First aid*. Mayo Clinic. https://www.mayoclinic.org/first-aid/first-aid-bruise/basics/art-20056663

Mayo Clinic Staff. (2024, May 1). *Cuts and scrapes: First aid*. Mayo Clinic. https://www.mayoclinic.org/first-aid/first-aid-cuts/basics/art-20056711

Mayo Clinic Staff. (2024, May 9). *Burns: First aid*. Mayo Clinic. https://www.mayoclinic.org/first-aid/first-aid-burns/basics/art-20056649

Smokey Bear. (n.d.). *How to maintain & extinguish your campfire*. https://smokeybear.com/en/prevention-how-tos/campfire-safety/how-to-maintain-and-extinguish-your-campfire

Yaghoobi, R., Kazerouni, A., & Kazerouni, O. (2013). *Evidence for clinical use of honey in wound healing as an anti-bacterial, anti-inflammatory, anti-oxidant and anti-viral agent: A review.* Jundishapur Journal of Natural Pharmaceutical Products, 8(3), 100–104. https://doi.org/10.17795/jjnpp-9487

Illustrations

This book includes a combination of original hand-drawn illustrations by the author's daughter, author-taken photographs, and design elements created using Canva under a commercial license.

To Ember —

Your passion for building shelters, foraging through the forest, and living wild and free inspired every page of this book.

You are the spark behind every adventure, every challenge, and every muddy footprint in these pages.

Keep that fire glowing, baby girl. The wilderness is lucky to have you.

Thank You!

Thank you for going on this epic adventure with us—we're so glad you took the challenge and stuck with it. Incredible!

If you loved *Wilderness Warriors*, check out a few of our other books, like:

Funniest Joke Book for Kids – for some good laughs around the campfire

Nature Journal for Kids – the perfect place to sketch, research, and record your wild discoveries

Taking Root: A Family Devotional – a different kind of adventure that strengthens faith and family with amazing discoveries and epic activities

We'd love it if you left a review on Amazon, Goodreads, wherever you bought your book—or all of the above!

As an independent publishing family, it's incredibly tough to get seen—and even tougher to be trusted.

Your support makes a big difference. Thank you a billion times over!

– The Hunters

Made in the USA
Monee, IL
13 July 2025

20845782R00193